Electrophysiology and Fish Behavior

The Author

Dr. K.P. Biswas, M.Sc, Ph.D, D.F.Sc. (Bombay), E.F. (West Germany), F.Z.S, F.A.B.S. (Kolkata), Former Joint Director Fisheries (L-I), Govt. of Odisha, Fishery Technologist, I.C.A.R. and Director of Fisheries, Andaman and Nicobar Islands, and at present Faculty Member, Marine Science Department, University of Calcutta and West Bengal University of Animal and Fishery Sciences is associated with fish, fisheries and marine sciences for more than fifty years. He has published twenty three books and 167 research and review papers on fish, fisheries, environmental and ocean sciences and fish processing technologies. He is a Fellow Member of Zoological Society, Indian Association of Biological Sciences, Kolkata. His latest book, " Conservation of Fishery Resources" was published in 2014.

Electrophysiology and Fish Behavior

K.P. Biswas

2015
Daya Publishing House®
A Division of
Astral International Pvt. Ltd.
New Delhi – 110 002

Cataloging in Publication Data–DK
Courtesy: D.K. Agencies (P) Ltd. <docinfo@dkagencies.com>
Biswas, K. P. (Kamakhya Pada), 1936- author.
Electrophysiology and fish behavior / K.P. Biswas.
　　pages　cm
　　Includes bibliographical references (pages　　) and index.
　　ISBN 978-93-5130-707-5 (International Edition)

　　1. Electrophysiology. 2. Fishes–Physiology–India. 3. Fishes–Behavior–India. 4. Electric fishing. I. Title.

　　DDC 571.17　　23

Published by　　　　: **Daya Publishing House®**
　　　　　　　　　　　　A Division of
　　　　　　　　　　　　Astral International Pvt. Ltd.
　　　　　　　　　　　　– ISO 9001:2008 Certified Company –
　　　　　　　　　　　　4760-61/23, Ansari Road, Darya Ganj
　　　　　　　　　　　　New Delhi-110 002
　　　　　　　　　　　　Ph. 011-43549197, 23278134
　　　　　　　　　　　　E-mail: info@astralint.com
　　　　　　　　　　　　Website: www.astralint.com

Laser Typesetting　　: **Classic Computer Services**, Delhi - 110 035

Printed at　　　　　　: **Thomson Press India Limited**

PRINTED IN INDIA

Acknowledgement

The author deeply acknowledge the help of Dr. N.A. Talwar for preparing the print out of hard copy of the manuscript.

K.P. Biswas

Preface

Electrofisheries mainly started for fishing with electric current from different quarters in different environments. Beginning with batteries and thereafter using induction coil to augment the voltage for catching fish, studies took place predominantly in Germany on the orientation and movement of fish exposed to direct current (Mach, 1875; Herman, 1885). In England Loeb and Maxwell (1896) demonstrated that the reactions were due to the stimulation of the motor neurons and were forced and not voluntary. Blasius and Schweizer (1893), however, discovered the phenomenon of "Galvanonarcosis", a state in which the fish appears to sleep with a relaxed body when it is facing the anode. Cotrastingly, a very strong stimulation was observed when the fish was facing the cathode.

In India, the author, while working as District Fishery Officer at Balasore, Odisha in 1960 was giving electric shocks to fishes in an aquarium in one evening, the then Director of Fisheiies, Odisha wanted to know what the author was doing. In the next year, the Director got a project through to work on electric fishing in deep water reservoirs by the author.

After conducting some laboratory experiments on the creation of underwater electric field and the response of Indian fishes in the field of increasing intensity, the author took up fieldtrials in ponds and reservoirs and designed an electrical seine net which effectively caught bottom living and mud dwelling and burrowing fish and prawns, normally escaped other fishing gears.

The author started publishing his findings in the journal of Fishery Technology and got acquinted with overseas experts in Germany. The author visited State Institute of Fisheries, North Rhine, Albaum and worked in the laboratory of Dr. H. W. Denzer for a period of six months on the behavior of fishes and crocodiles in electrical field with modern and sophisticated equipments. On his return to the country, the author started working on Indian fishes, prawns and crabs and published a number of papers on his research results.

The electrical fishing work mainly involve two parts. Firstly, how the aquatic animals (exploitable living resources) respond to the underwater electric field and orient themselves in the field with increasing intensity, and how these animals can be influenced by the electric field ? Secondly the problem is of technical electricity to create the field with greater area of influence by different current types, frequencies and current forms.

For the first problem, again, it is to be decided, what type of influence is desired of the animals, attracting to catch them, or frighten and block them or stunning them for instantaneous pre-rigor ? Once this is decided, it is the work of technical electricity to create the field of appropriate intensity with proper frequency and wave form.

Electric fishes, themselves have shown the way to man. With the discovery of Adanson (1757) about the similarity between electric discharge produced by a fish (*Malapterurus*) with that of Leydon jar and of the powerful discharge of *Torpedo* (Walsh, 1973) and *Electrophorus electricus* (Williamson, 1975), the catching and guiding of aquatic animals with the aid of underwater electric field has aroused a special interest among research workers.

The reaction of animals observed by different workers depended only partly on the nature of applied current. The orientation and movement of fish in a direct current were described by Mach (1875) and Hermann (1885); while in 1894 Hermann and Matthias reported that these reactions were abolished by destruction of the spinal cord.

The animal behavior in an electric field varies according to the taxonomic group. The galvanotaxis is cathodic in Annelidae. These differences reflect variation in the reflex activity and nervous system of the animals (Blancheteau *et al.*, 1961). For the worm the reflex possibility is quasi-nil, thus it moves towards the cathode, and it is only stimulated when it faces the electrode.

Differences have also been observed between species of the same animal class, particularly fish. These differences primarily relate to the initial response to weak electric fields caused by the brain reflex. In general, each species exhibits its behavioral fright response, that is, a pelagic fish will swim rapidly, a benthic species will burry itself in the substrate and a cryptic species will hide itself. As the field strength increases these behavioral responses are over-riden because the usual reflex cannot overcome the effect of the stronger electric stimulus.

The author in 1974 studied on the effect of underwater electrical field on three different varieties of Indian fish, namely, a barb (*Puntius ticto*), a cat fish (*Heteropneustis fossilis*) and a cichilid (*Tilapia mossumbica*) and obtained M.Sc. degree from University of Bombay.

Further studies were undertaken by him on other ten varieties of Indian fishes and fresh water prawn for their anatomical and behavioral pecularities.The author also worked to determine the effect of electric shocks on the development of fish embryos, young larvae and their orientation in the electric field. Efforts have also been made to identify and describe the responses of animals in relation to field intensity and nature of current. Responses were measured as a function of the maximum intensity of the electric field produced in the water, electrical conductivity of the water, types of current, frequency of impulses and the length of the subject animal along with its orientation in the field. The effect of electric shocks on fish muscle contraction and on the heart beat was also determined along with electrical resistance of some of the subject species. The author also studied harmful effects of electric shocks on the fertilized eggs and developing embryos to find out the lethal limits of the exposure. He further studied the galvanotropic responses of the organisms in their developing stages and growth, besides the threshold limits of current intensities to their death point. For all these works, the author was awarded Ph.D. degree of University of Bombay in 1977.

Kamakhya Pada Biswas

K.P. Biswas

Contents

Muscular System–Electrotonus : DC Action on the Body Cells–Mechanism behind the Reactions–Rushton's Law–Mechanism of Anodic Galvanotaxis–Mechanism of Muscular Tetanus–Responses in Pulsed Current Field–Physiology of Reactions in Pulsed Current Field–Mechanism of Anodic Electrotaxis–Mechanism of Tetanus–Alternating Current–Comparison between DC, Pulsed Current and AC Reactions.

Introduction

Visible signs of behavior of fish subjected to an electric field, that is, their physical reactions attract a good deal of attention in controlling fish movement for capture and management. The definite sequence of reactions with an increase of field strength was observed. Investigations were carried out with different types of current, various species and length of fish, their physiological condition and environment parameters (temperature and water conductivity). Nearly all these experiments were held under laboratory conditions and what is of particular significance here is in homogeneous electric field. More important still in experiments aimed at the determining voltage gradients, current densities or head-to-tail voltage necessary to produce "the first reaction" or "electro-narcosis", it appeared necessary for getting uniform results, to place fish exactly along the lines of current flow, that is, perpendicular to the equipotential lines. An electric field affects more strongly when head-to-tail voltage is higher. If a fish find itself just along the equipotential line, that voltage equals zero, and the fish hardly feels any electrical stimulus, even if the electrical field is strong.

Pure water is a poor conductor of electricity. Current in solution dissociates molecules of the substance, known as, electrolytic dissociation. In solution, the molecules of acids, alkalies and salts dissociate into the positive and negative ions. The conduction of current through an electrolyte is a convection effect, the charges being carried to the electrodes by the ions and thereby differ from the ordinary metallic conduction, where it involves transfer of matter and is accompanied by chemical change.

If a homogeneous field is produced, the intensity of the field is known as current density and is expressed as delta which means ten to the power minus six ampere current flowing through water column in one square millimeter.

In an equivalent bath, where conductivity of water equals to the conductivity of fish body, no visible reaction is observed and the fish taps off the potential lines corresponding to its length.

When the conductivity of water is lower than the fish body, all the potential lines are directed towards the fish body with better conductivity and the fish is thus satisfactorily influenced by electricity. The body voltage amounts to 2 volts as compared to 5.3 volts in the equivalent bath.

The conductivity of sea water is better than the conductivity of fish body. The potential lines are distorted by the fish owing to its lesser conductivity. The body voltage increases to 8 volts as compared to 2 volts in fresh water and 5.3 volts in equivalent bath.

In order to produce a fishing effect at a certain distance from the electrode, a current of such strength must flow that a fish of length (L) can receive a body voltage (G) for the galvanotaxis. The current density (I) that must exist around the fish to effect these reactions can be computed from the length of fish, the body voltage and the specific resistance of sea water.

The striking phenomenon of stunning fish for capture with electric current, attracted serious attention of many people after the first World War. A British patent (No. 2644) for electric fishing was granted to Isham Baggs as early as 1863, and two others to Takahashi in 1895 (No. 2551 and 2555).

The effectiveness of the equipment's output energy is sometimes reduced drastically by the environmental or biological factors. The resistive effects of water do not alter pulse frequency or duration, but depresses the electric energy input. Some small adjustments in output power can be made to reduce the erratic action and escape of fish.

According to the varying effect of electrical current on the fish, a distinction can be made between fishing gear with;

(a) Attracting anodic effect;

(b) Frightening or blocking effect; and

(c) Narcotizing or killing effect.

All these three groups are used in fresh water and sea fishing. For attracting (anodic) effects in fishing at sea, an impulse generator, containing a condenser, which is loaded relative slowly by a direct current generator is discharged within a very short time over a throttle and passes the concentrated energy over the electrodes into the water to obtain high peak voltages required. Adjustment of loading and discharging is done by means of an electronic governor, whereby the discharge current is switched over two parallely connected ignitrons. The rate of pulses can be varied according to species of fish, from 3 to 60 impulses per second. The length of the impulses can be changed in the ratio of about 1:3.

The blocking or driving effect is used as barriers for controlling the migration of Chinese crabs, considered as unwanted pest to shore installations and gear. The blocking chain used consists of four iron cables of 100 square mm in cross section with a copper core of 6 mm in cross section. The cables are so arranged that three electric fields develop; a small central field of 4 cm width and two outer fields of 10 cm in width end. The crabs entering the electrical field of interrupted current are either killed or they throw off their limbs (ecdysis) when the circuit is connected and the animal can not continue their migration.

To prevent fish being driven into turbines and pumps of hydro-electric dams, the electrical fish barriers are installed with a controlling device connected to 220 to 280 volts A.C. and one or several chains of electrodes are hung in the river or water area to be fenced, considering the water depth and bottom with particular importance to the speed of water flow, fluctuation of water level, conductivity and water temperature.

Replacing the conventional methods for diverting salmon employing mechanical barriers (which usually impede stream flow and costly), the use of electrical field created by 110 volts A.C. was found to be the most effective method for the diversion of upstream migrants. The voltage gradient should vary 0.3 to 0.7 volts/2.5 cm.

The fishing industry became interested in stunning and killing for collecting fish and prawn from holding pond for retail trade, on the one hand, while the sea fishermen are interested to kill or paralyze small pelagic fishis caught in seine net to prevent loss of fish scales and in the dip net or on board and for delaying the onset and disappearance of rigor mortis, thereby improving the fish flesh quality.

Fishing gears in shape of weirs and traps installed in rivers and estuary use electrical influence on fish. Salmon entering the range of the guiding bar are driven by the effect of galvanotaxis produced by the installed anode and by the water current towards the trap net. The stunned fish are then swept away into the catching box by water current.

Five percent of the hooked tuna generally escape from long line. Also during long struggle on the hook (up to 30 minutes), the lactic acid content in fish muscle increase, detoriating the keeping quality of the meat. The electrical tuna long line enable the tuna to be paralyzed immediately it is hooked and pulling the line, which causes a contact and releasing the flow of current to fish body. The electrical impulses are transmitted over a copper wire to the hook, pass into the fish body and then to the sea water.

Electrical current has long been used in taking and killing whales. Owing to the effect of the electric shock produced by the pulsating current, the whale is paralyzed as soon as the harpoon hits it. Alternating current of 220 volts is used. The ship's hull act as zero conductor and the forerunner of the harpoon gun, together with harpoon head is used to conduct the potential. It is also thought that the pulsating current provides an opportunity to carry out electrical whaling from helicopters.

The human body is adversely affected by the electric current. Intensity of current passing through the body depends on the voltage existing at the moment of contact and electrical resistance of the body. Current more than 50 mA are dangerous to human life if the influence exists for 0.2 second, which means the dangerous contact voltage is 40 volts.

Fishing by skilled fishermen under the guidance of an expert and use of electrical fishing gear with safety and protective devices can prevent electrical accidents.

Chapter 1

The History of Electric Fishing

Fishing with electricity developed from different origins in different environments, and thus has a confusing history. In this it resembles most practical types of activity which depend on the exploitation of a physical principle.

The first definite record is the patent granted to Mr Isham Baggs, living in Islington, London, in 1863. This is notable in revealing a thoroughly practical approach which must have been the outcome of considerable experience; he advised the use of polarizing spectacles to help the operator. At that date he was handicapped by the lack of any powerful electric apparatus, and relied on batteries. Like many after him, he augmented the voltage by using an induction coil, and thus pioneered the exploitation of high-powered gear.

Following Isham Bagg's work, there was no practical development of the process, but studies took place, predominantly in Germany, on the orientation and movement of fish exposed to direct current (Mach, 1875; Herman, 1885). In England, Loeb and Maxwell (1896) demonstrated that these reactions were due to stimulation of the motor neurons and were forced, not voluntary. Blasius and Schweitzer (1893), however discovered the phenomenon of *galvanonarcosis*, a state in which the fish appears to sleep with a relaxed body when it is facing the anode. This contrasted with the strong stimulation they observed when the fish was facing the cathode.

It seems that practical electric fishing developed mainly in Germany with Holzer (1932) and the support of physiological research of Scheminzky (1924). However, early in the 20th century there are accounts of Japanese fishermen using induction coils to drive eels out of their burrows into a net (Takahashi, 1895). Also a patent claim by Larssen in 1912 covered the use of electricity to catch a variety of aquatic creatures from fish to seals, suggesting that the claimant recognized the possible potential.

The earliest reliable record of an electric fish screen was in 1917 in the United States when Burkey took out the first (Patent No. 1,269,380) of many patents. Systematic work on screens, however, really dates from the masterly analysis of Mac Millan (1928) and Tauti (1931) which enunciated the principles of aquatic fields and their logical use as fish screens.

After the last war, the development of electric fishing and electric screens continued in Germany, primarily with Denzer (1956) and Halsband (1959) as the driving forces. At the same time electric fishing engines were being produced commercially and this opened up a whole new dimension. During this time the control of fish behavior was quite familiar, in Germany by direct current in tanks and in America by alternating current, and a series of papers appeared, mainly in Germany(namely, Scheminzky, 1922, 1924, 1934; Scheminzky and Kollesperger, 1938).

Post-war American was initially concerned itself with the construction of portable apparatus capable of being used by a single individual and suitable for locations difficult to access. More recently, however, design shifted towards large fishing devices based on what can only be considered as fish-screens carried on special rafts moved by propulsive motors in open water. The formerly ubiquitous employment of alternating current was replaced by pulsed direct current provided by electronic devices from an alternating supply. This had an effect on the thinking underlying the whole process, introducing a degree of formality into the procedure, which was previously unnecessary, as a precaution against fatal accidents which had by now occurred. The days of experimental apparatus has passed. In general, fishing machines had to be commercially constructed to approved standards.

The situation in Great Britain after 1945 developed from both German and American routes. The fishery authority of eastern England, where the rivers were mainly sluggish and inhabited by coarse fish, followed along American lines and used alternating current. Simultaneously, in the fast flowing, high conductivity chalk streams of southern England, experiments took place with direct current to facilitate the removal of coarse fish from these trout waters.

During the time considerable research and development was taking place elsewhere. For example, in the Soviet Union, Strakhov and Nusenbaum (1959) were developing electric screens and Schentiakov (1960) electric trawling in lakes. In North America, Applegate *et.al* (1952) were producing an efficient electric screen against the invasion of sea lampreys into the Great Lakes and in New Zealand, Burnet (1959) was experimenting with electric fishing.

In 1957, in Hamburg, a first synthesis of electric fishing problems was resolved in the first FAO International Fishing Gear Congress in 1963, although discussion on electric fishing was minimal. Thereafter, Blanchteau *et.al.* (1961) in France and Nusenbaum and Faleeva (1961) in the Soviet Union, introduced the elements of comprehensive theory of electric fishing, based on physiological principles.

In 1965, the European Inland Fisheries Advisory Committee (EIFAC) arranged a meeting, in Bairritz, of workers in the electric fishing field. This working party received papers on various aspects of the technique, which were subsequently embodied in a book *Fishing with Electricity* (Vibert, 1967a), representing its report to the Fourth Session of EIFAC at Belgrade in 1966.

In 1973, the working party reconvened in Gysisko, Poland, to compare the efficiency of portable apparatus used by participants at that time (EIFAC, 1973), and elaborate tests on the fatigue and mortality induced by the gear. The results were published by Chmielewsky (1973a,b). During this period a book on the general principles of electric fishing was produced by Sternin *et.al.* (1972) and Halsband and Halsband (1984) updated their book on the subject.

More recent developments were reported to the EIFAC Symposium on Fish Population Studies held at Aviemore, Scotland in 1974. This provided descriptions of new designs in fishing apparatus, and the instruments for the detection and counting of migrant fish passing a census point in a channel. In 1975, Lamarque, in conjunction with the FOOD and Agriculture Organization (FAO) completed a comprehensive study on electric fishing in tropical waters which elaborated on the theory of Blanchteau *et. al.*(1961), allowing the choice of the best current for catching fish and opening up the practice of fishing in sea water and low conductivity water (Lamarque, 1977b).

Under present conditions, technical advances in electronics have now made it possible to design and construct light-weight fishing apparatus with virtually any current characteristic, thus increasing the potential for future development.

Electric Fishes

Royce, Smith and Hartt (1968) have revived the suggestion that aquatic animals might use the weak electric fields generated in the ocean by water currents moving through the geomagnetic field for orientation or navigation. They suggested that Pacific salmon (*Oncorhynchus*), which migrate along ocean currents, may determine the direction of the water currents by means of this geoelectric field. Others have made similar hypotheses (Regnart, 1932; Waterman, 1959; Lissmann, 1958). Deelder (1952) offered electric field detection as one of the possible means by which migrating elvers of the European eel (*Anguilla anguilla*) orient themselves in tidal streams.

Several groups of essentially non-migratory fish have been shown to have sensitivity to very weak electric fields, for example, the weakly electric fish (Gymnotids, Mormyridae) to 0.03 micro-volt per cm, sharks (*Scyliorhinus canicula*) and skates (*Raja clavata*) to 0.01 micro-volt per cm, and bullheads (*Ictalurus nebulosus*) to 30 micro-volt per cm. But determination of the electric current density to which the fish were exposed is not possible, since a non-uniform field was applied to known receptor organs, and the resistivity of the water was not stated.

Studies on the responses of American eels (*Anguilla rostrata*) exposed to uniform weak electric fields reveal, that the eel's sensitivity is well within the range of naturally occurring oceanic electric fields. In certain study sections of the Gulf stream, the predicted values were up to 0.46 micro-volt per cm. Electric fields are at a maximum in the Gulf Stream, but surface potential gradients of at least 0.10 micro-volt per cm also occur, for example, in the Labrador, West Greenland, North Atlantic, and Antilles currents. All of these are almost certainly involved in migration routes of American and European eels.

While demonstration of a sensitivity does not prove its use in orientation, the fact that the eel is sensitive to perpendicular fields, but not parallel fields, provides a mechanism by which water current direction can be determined. Interaction of moving water with the vertical component of the geomagnetic field would generate a potential gradient perpendicular to the body axis, if the eelwere oriented upstream or downstream.

The potential gradient would be parallel to the body, if the eel were oriented across the water current. By simply aligning the body to sense the electric current the eel could remain oriented to the water current. If it can also sense polarity, it could discriminate upstream from downstream. The horizontal component of the geomagnetic field interacting with moving water produces vertical fields, which would provide no directional information to fish. However,most migratory fish, such as salmon and eels, have routes distinctly north of the equatorial region, where the vertical component is the major portion of the total geomagnetic field.

Since it is sensitive perpendicular but not parallel to the body, an eel in a natural water current system would experience a rapid change in electric field when it turned its body from side to side as in its swimming movements.

The sense organs transform energy of various kinds, heat and light, mechanical energy and chemical energy into nerve impulses. Since human organisms is sensitive to the same kinds of energy, man can to some extent visualize the world as it appears to other living things.

Gymnarchus niloticus, a fish of African origin can perceive an environmental stimuli through its electrical sense organ. Its grace of swimming can impress a casual observer. It does not lash its tail from side to side, as most other fishes do, but keeps its spine straight. An undulating fin along its back propels its body through the water forward or backward with equal ease. *Gymnarchus* can maintain its rigid posture even when turning with complex wave forms running hither and thither over different regions of the dorsal fin at one and the same time.

Close observations reveal that the movements are executed with great precision. On darting after the small fish on which it feeds, it never bumps into the walls of its tank and it clearly takes evasive action at some distance from obstacles placed on its path. Such maneuvers are not surprising in a fish swimming forward, but *Gymnarchus* performs them equally well swimming backward. As a matter of fact it should be handicapped even when it is moving forward, its rather degenerated eyes seem to react only to excessively bright light.

Another unusual aspect of this fish, is its tail, a slender, pointed process bare of any fin (naked tail). The tail was first dissected in 1847 and was found tissue resembling a small electric organ, consisting of four thin spindles running up each side to somewhere beyond the middle of the body.

Such small electric organs have been an enigma for a long time like the powerful electric organs of electric eel and some other fishes, they are derived from muscle tissue. Apparently in the course of evolution the tissue lost its power to contract and became specialized in various ways to produce electric discharges. In powerful electric fishes this adaptation serves to deter predators and to paralyze prey. But the powerful electric organs must have evolved from weak ones. The original swimming muscles would therefore seem to have possessed or have acquired at some stage a subsidiary electric function that had survival value.

On placing a new object, the *Gymnarchus* would approach it with caution, making what appeared to be exploratory movements with the tip of its tail. The electric organ in the tail might be a detecting mechanism. On putting a pair of electrodes, connected to an amplifier and an oscilloscope, instead of sporadic discharges coordinated with the swimming or exploratory motion of the animal, a continuous stream of electric discharges at a constant frequency of about 300 per second, waxing and waning in amplitude as the fish changed its position in relation to the stationary electrodes were observed. Even when the fish was completely motionless, the electric activity remain unchanged.

This was the first observation of an electric fish behaving in such a manner. Two other kinds, one a mormyrid relative of *Gymnarchus* and the other a gymnotid, a small fresh water South American relative of electric eel, belonging to a group of fish rather far removed from *Gymnarchus* and the mormyrids were found to emit uninterrupted stream of weak discharges.

It has been known for some time that the electric eel generates not only strong discharges, but also irregular series of weaker discharges. Various functions had been ascribed to these weak discharges of the eel. They might serve in navigation, postulating that the eel somehow measured the time delay between the output of a pulse and its reflection from an object. The idea was untenable on physical and physiological ground. The eel does not produce electro-magnetic waves, if it did, they would travel too fast to be timed at the close range at which such a mechanism might be useful, and in any case they would hardly penetrate water. Electric current, which the eel does produce, is not reflected from the objects in the surrounding environment.

Observation of *Gymnarchus* suggested another mechanism. During each discharge, the tip of its tail becomes momentarily negative with respect to head. The electric current may thus be pictured as spreading out into the surrounding water in the pattern of lines that describes a dipole field (Figure 1.1)

The exact configuration of this electric field depends on the conductivity of the water and on the distortions introduced in the field by objects with electrical conductivity different from that of the water. In a large volume of water containing no objects, the field is symmetrical. When objects are present, the lines of current will converge on those that have better conductivity and diverge from the poor conductors (Figure 1.2).

Such objects alter the distribution of electric potential over the surface of the fish. If the fish could register the changes, it would have a means of detecting the objects. *Gymnarchus* was sensitive to extremely small external electrical disturbances. It responded violently when a small magnet or an electrified insulator was moved near

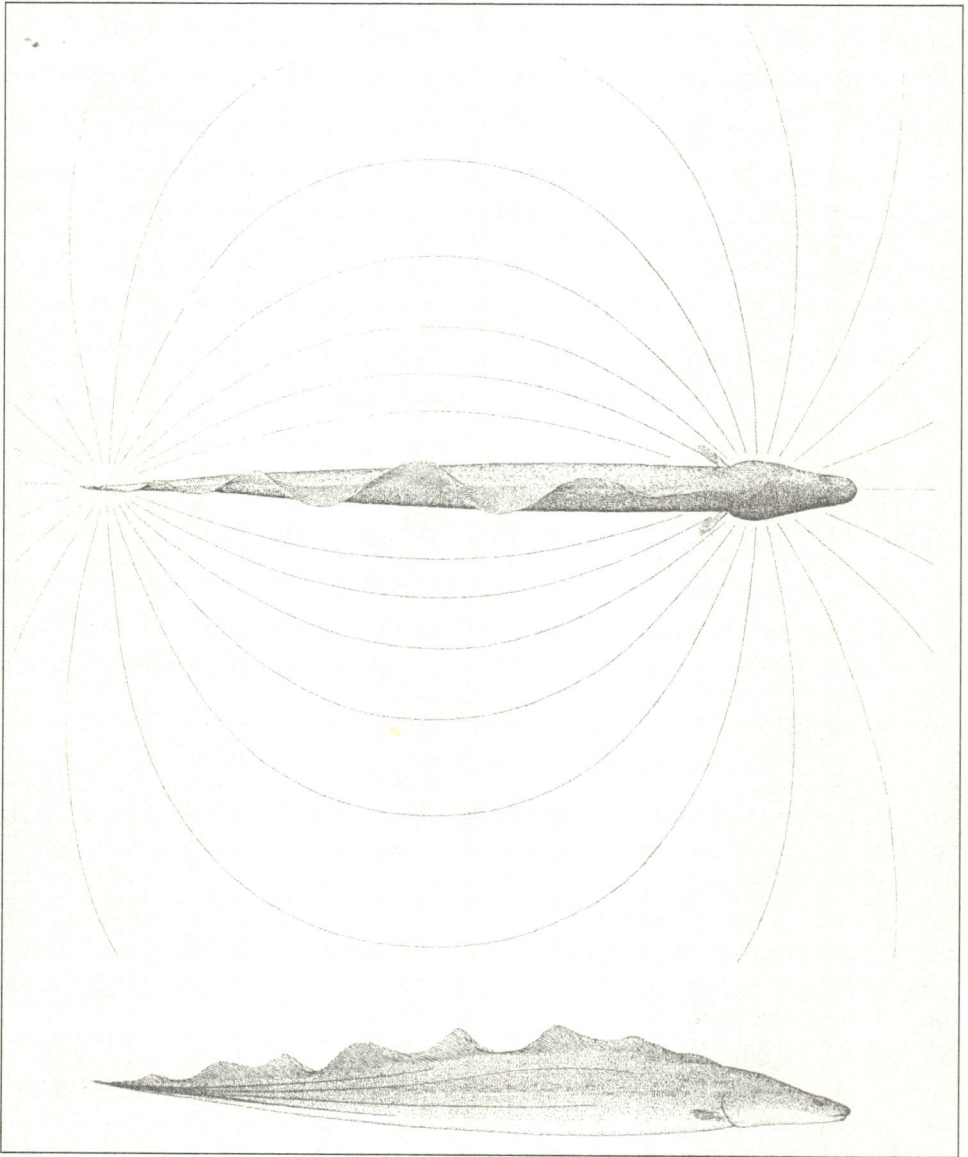

Figure 1.1: Electrical field of *Gymnarchus* **and location of electric generating organs are shown. Each electric discharge from organs in the rear portion of the body makes tail negative with respect to head. Most of the electric sensory pores or organs are in the head region. Undisturbed electric field resembles a dipole field as shown but is more complex. The fish responds to change in the distribution of electric potential over the surface of its body. The conductivity of the objects affects distribution of potential.**

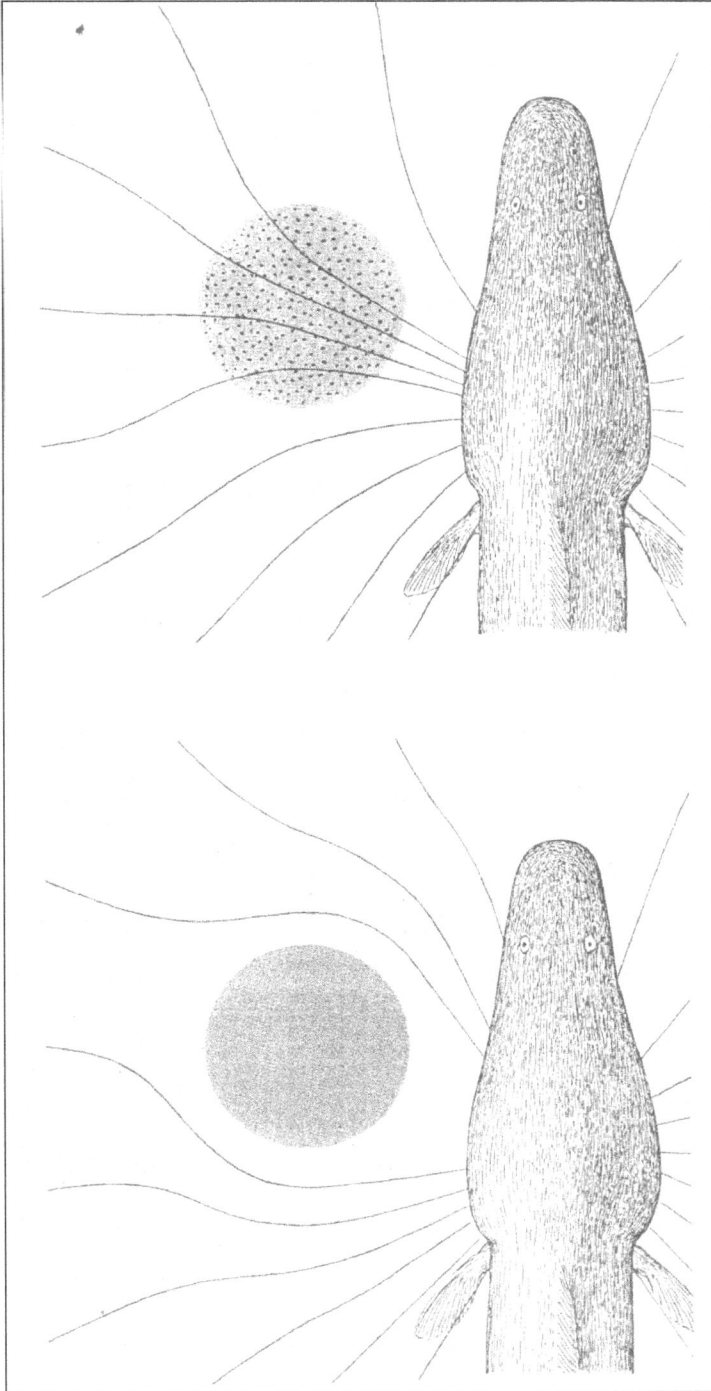

Figure 1.2: Objects in electric field of *Gymnarchus* distort the lines of current flow. The lines diverge from a poor conductor (left) and converge towards a good conductor (right). Sensory pores in the head region detect the effect and inform the fish about the object.

the aquarium. The electrical field produced by such object is very small, in the range of fractions of a millionth of one volt per centimeter.

A series of experiments seemed to establish beyond reasonable doubt the *Gymnarchus* can detect objects by an electrical mechanism.

While seeking for the possible channels through which the electrical information may reach the brain, it was found that the tissues and fluids of a fresh water fish are relatively good electrical conductors enclosed in a skin that conducts poorly. The skin of *Gymnarchus* and of many mormyrids is exceptionally thick with layers of plate-like cells sometimes arrayed in a remarkable hexagonal pattern (Figure 1.3).

It can therefore be assumed that natural selection has provided these fishes with better than average exterior insulation.

On and around the head region, the skin is closely perforated. The pores lead into tubes often filled with a jelly like substance or loose aggregation of cells. If this

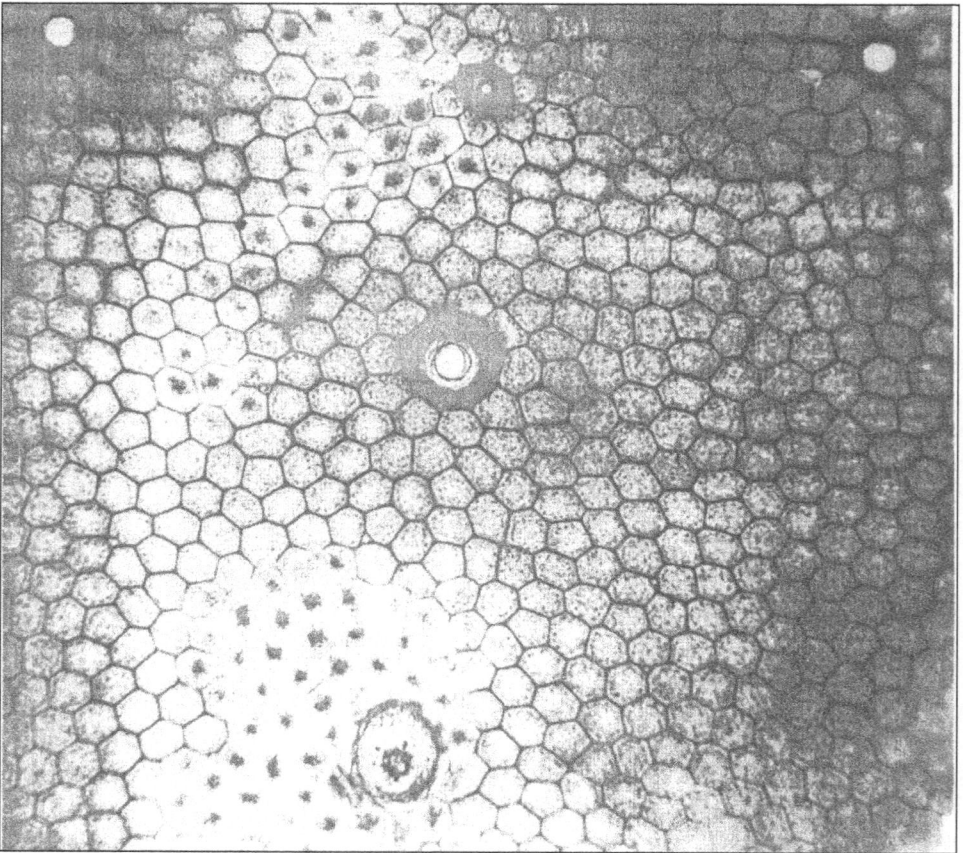

Figure 1.3: Skin of Mormyrid is made up of many layers of plate-like cells having remarkable hexagonal structure. The pores contain tubes leading to electric sense organ.

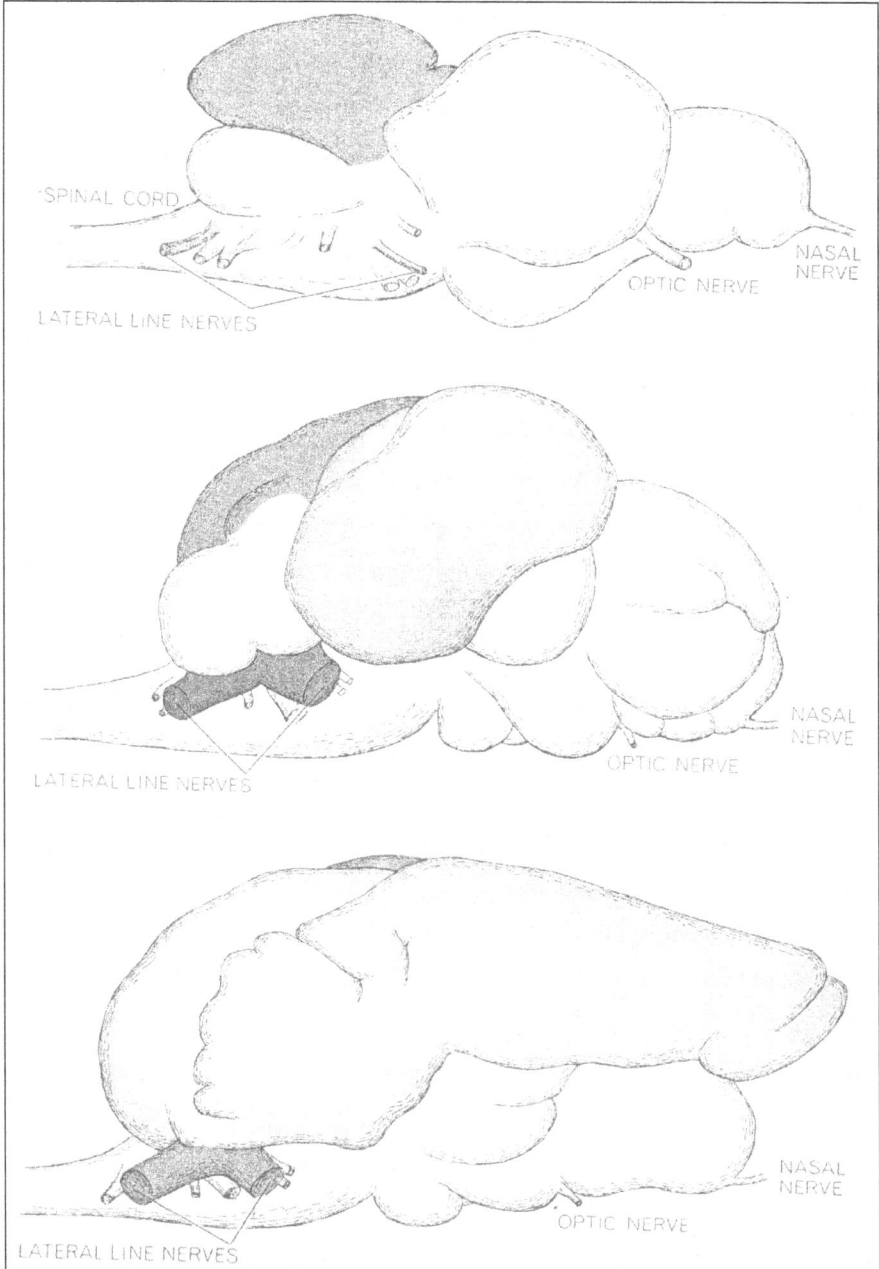

Figure 1.4: Brain and nerve adaptation of electric fish are readily apparent. Brain of a typical non-electric fish (top) has prominent cerebellum. Regions associated with electric sense are quite large in *Gymnarchus* (middle) and even larger in the mormyrid (bottom). Lateral line nerves of electric fishes are larger, nerves of nose and eyes smaller.

jelly is a good electrical conductor, the arrangement would suggest that the lines of electric current from the water into the body of the fish are made to converge at these pores as if focused by a lens. Each jelly filled tube widens at the base into a small round capsule that contain a group of cells, known as " multi-cellular glands", "mormuromasts" and "snout organs", which is believed to be electric sense organs.

The supporting evidence appears fairly strong. The structures in the capsule at the base of a tube receive sensory nerve fibers that unite to form the stoutest of all the nerves leading into the brain. Electrical recording of the impulse traffic in such nerves has shown that they lead away from organs highly sensitive to electric stimuli. The brain centers into which these nerves run are remarkably large and complex in *Gymnarchus* and in some mormyrids, they completely cover the remaining portion of the brain (Figure 1.4).

Except for the electric eel, all species of gymnotids investigated so far emit continuous electric pulses. They are also highly sensitive to electric fields. Dissection of these fishes reveal, that expected histological counterparts of the structures found in the mormyrids; similar sense organs embedded in a similar skin, and the corresponding regions of the brain much enlarged.

Skates also have a weak electric organ in the tail. They are cartilaginous fishes, far removed from the family line of bony fishes and live in the sea, which conducts electricity much better than fresh water. Yet skates posses sense organs known as ampullae of Lorenzini, that consists of long jelly filled tubes opening to the water in one end and terminating in a sensory vesicle at the other. These organs respond to very delicate electrical stimulation.

Gymnarchus, the gymnotids and skates all share one obvious feature; they swim in an unusual way. *Gymnarchus* swims with the aid of a fin on its back, the gymnotids have a similar fin on their underside; skates swim with pectoral fins stuck outside ways like wings (Figure 1.5).

They all keep the spine rigid as they move. Such deviations from the basic fish plan could be attributed to an accident of nature.

In the case of the electric sense organ of a fish the stimulus energy is provided by the discharges of the animal's electric organ. *Gymnarchus* discharges at the rate of 300 pulses per second. A change in the amplitude, not the rate of these pulses caused by the presence of an object in the field, constitutes the effective stimulus at the sense organ. Assuming that the reception of a single discharge of small amplitude excites one impulse in a sensory nerve, a discharge of larger amplitude that excited two impulses would probably reach and exceed the upper limit at which the nerve can generate impulses, since the nerve would now be firing 600 times a second (twice the rate of discharge of the electric organ). This would leave no room to convey information about gradual changes in the amplitude of incoming stimuli. Moreover, the electric organs of some gymnotids discharge at a much higher rate. 1000 pulses per second have been recorded.

It was found that the fish is in fact as sensitive as high frequency pulses of short duration as it is to low frequency pulses of identical voltage, but correspondingly

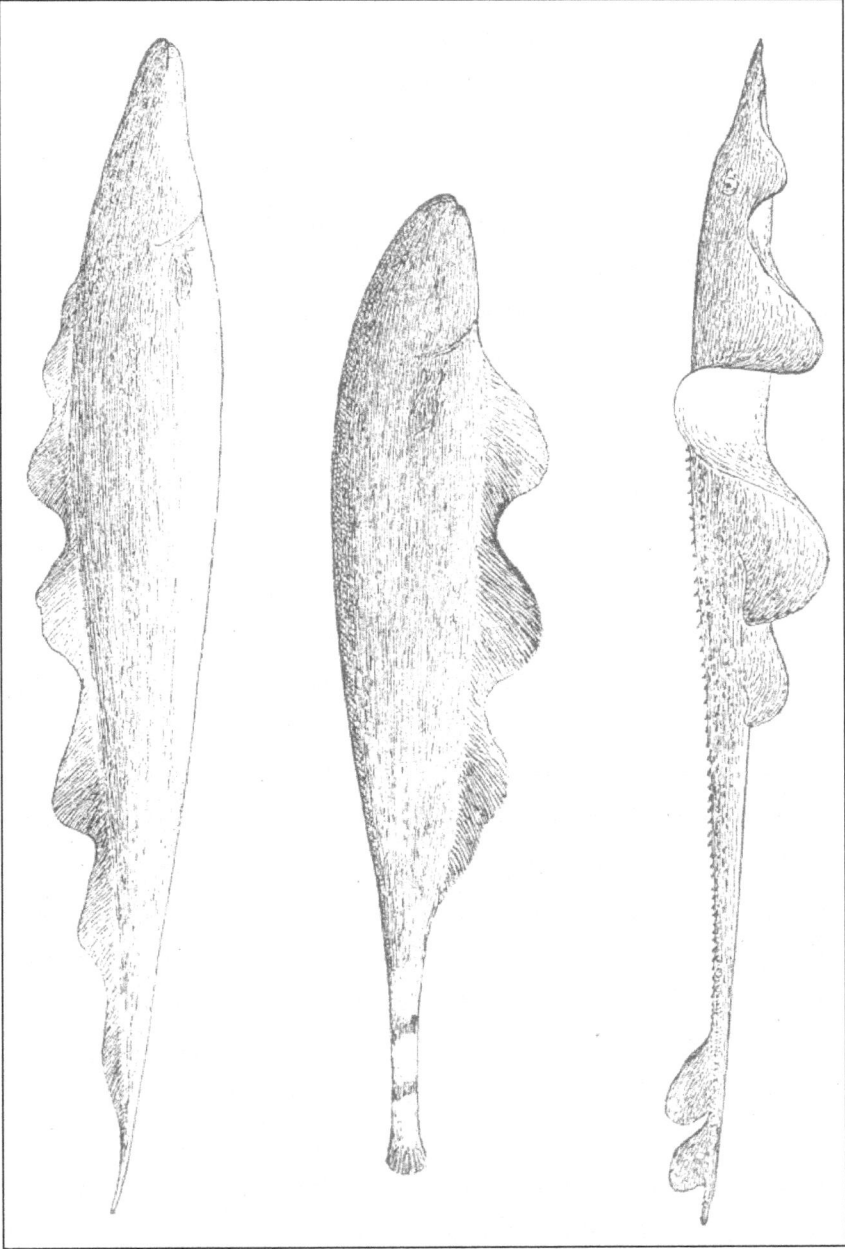

Figure 1.5(i): Unusual fins, characterize *Gymnarchus* (top), a gymnotid from South America (middle) and sea dwelling skate (bottom). All swim with spine rigid, probably in order to keep electric generating and detecting organs aligned. *Gymnarchus* is propelled by undulating dorsal fin, gymnotid by similar fin underneath and skate by lateral fins resembling wings.

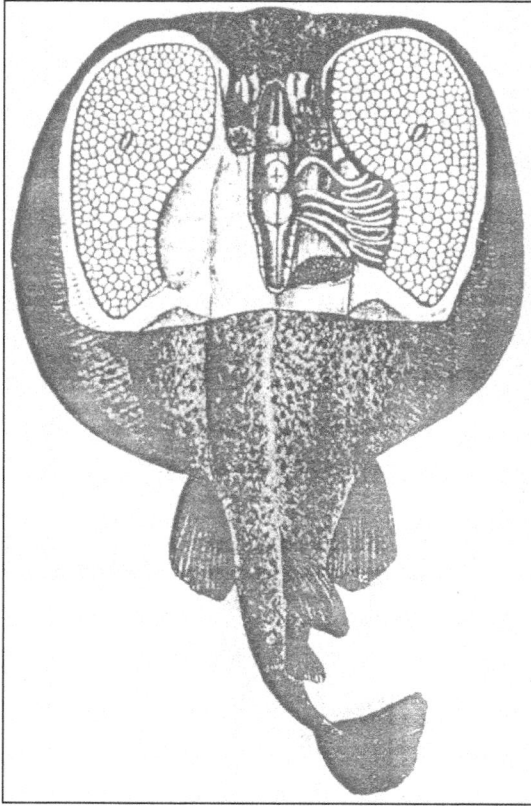

Figure 1.5(ii): *Torpedo marmorata* showing electric organ.

Figure 1.5(iii): *Malapterurus electricus* electric cat fish.

longer duration. For any given pulse train reduction in voltage could be compensated either by an increase in frequency of the stimulus or an increase in the duration of the pulse. Conversely, reduction in the frequency required an increase in the voltage or in the duration of the pulse to reach the threshold. The threshold would therefore appear to be determined by the product of voltage, times duration, times frequency.

Since the frequency and the duration of discharges are fixed by the output of electric organ, the critical variable at the sensory organ is voltage. Threshold determination of the fish's response to single pulse compared with quantitative data on its response to trains of pulses made it possible to calculate the time over which the fish averages out the necessarily blurred information carried within a single discharge of its own. This time proved to be 25 milliseconds, sufficient for electric organ to emit 7 or 8 discharges.

It has been found that *Gymnarchus* can respond to a continuous direct current electric stimulus of about 0.15 microvolt per centimeter, a value that agrees reasonably well with calculated sensitivity required to recognize a glass rod 2 mm in diameter. This means that an individual sense organ should be able to convey information about a current as 0.003 micro ampere. Extended over the integration time of 25 milliseconds, this tiny current corresponds to a movement of some 1000 univalent or singly charged ions.

The intimate mechanism of the single sensory cell of these organs is still a complete mystery. In structure the sense organs differ somewhat from species to species and different types are also found in an individual fish. The fine structure of the sensory cells, their nerves and associated elements show many interesting details. Along specialized areas of the boundary between the sensory cells and the nerve fiber there are sites of intimate contact where the sensory cells bulges into the fiber. A dense streak extends from the cell into this bulge, and the vesicles alongside it seem to penetrate the intercellular space. The integrating system of the sensory cell may be here.

These findings, however, apply only to *Gymnarchus* and to about half of the species of gymnotids investigated to date. The electric organs of these fishes emit pulses of constant frequency. In the other gymnotids and all the mormyrids the discharge frequency changes with the state of excitation of the fish. There is therefore no constant mean value of current transmitted in a unit of time; the integration of information in these species may be carried out in the brain. Nevertheless, it is interesting that both types of sensory system should have evolved independently in the two different families, one in Africa and one in South America.

The experiments with *Gymnarchus* which indicate no information is carried by the pulse nature of the discharges, leave with a still unsolved problem. If the pulses are smoothed out, it is difficult to see how any one fish can receive information in its own frequency range without interference from its neighbours. The potentiall significant findings of Akira Watanabe and Kimihisa Takeda at the University of Tokyo indicate that the gymnotids respond to electric oscillations close in frequency to their own by shifting their frequency away from the applied frequency. Two fish might thus react to each other's presence.

For reasons that are associated with the evolutionary origin of their electric sense to the electric fishes are elusive subjects for study in the field. In their natural habitat, a *Gymnarchus*, a mormyrid or a gymnotid can be rarely seen in the turbid waters in which they live. While such waters must have favored the evolution of an electric sense, it could not have been the only factor. The same water contain large number of other fishes that apparently have no electric organs.

Although electric fishes can not be seen in their natural habitat, it is still possible to detect and follow them by picking up their discharges from the water. In South America the gymnotids are found as alive during the night. Darkness and turbidity of the water offer a good protection to these fishes, which rely on their eyes only for the knowledge that it is day or night. At night most of the predatory fishes, which have well developed eyes sleep on the bottom of rivers, ponds and lakes. Early in the morning before the predators wake up, the gymnotids return from their nightly excursions and occupy in accessible hiding places, where they often collect in large numbers. In the rocks and vegetation along the shore the ticking, rattling, humming and whistling can be heard in bewildering profusions when the electrodes connected to the loud speaker. With a little practice one can begin to distinguish the various species by their sounds.

Life in this highly competitive environment, the advantages the electric sense confers on these fishes by evolving their curiously specialized sense organs, skin, brain, electric organs and peculiar mode of swimming. These special features must have originated, however, from ordinary fishes in which the characteristics of the special features are found in their primitive state; the electric organs as locomotive muscles and the sense organs as mechano-receptors along the lateral line of the body that signal displacement of water. Somewhere there must be intermediate forms in which the contraction of a muscle with its accompanying change in electric potential, interacts with these sense organs. For survival it may be important to be able to distinguish water movements caused by animate or inanimate objects. This may have started the evolutionary trend toward an electric sense. From non-electric fishes signals having many characteristics of the discharges of electric fishes could be picked up. Sense organ that appear to be structurally intermediate between ordinary lateral line receptors and electro-receptors. Fishes that have both of these characteristics are also electrically very sensitive.

In *Torpedo marmorate* (electric ray), the electric organ are found to be large covering at least one-third to one fourth of the anterior region of the body located on both sides of the circular disc. These electric organs have large number of muscular plates, each of which act as a capacitor to collect 0.1 volt and emit short pulses of current discharge. Electric eel belonging to Gymnotid family is able to produce 400 impulses per second with peak voltage of 300 to 600 volts. *Malapterurus electricus* of family Siluridae inhabitant of Nile and Africa can produce electric pulses of 350 volts at an impulse frequency of 280 per second.

Chapter 2

Electrical Parameters

For the best results in electro-fisheries, it is necessary to have a knowledge of electrical parameters. This is essential for the variable fishing requirements and the widely varying conditions under which such systems are required to operate. Physical variables such as electrical conductivity of water, clarity and temperature as well as fishing requirements such as the species desired and methods of capture, all affect the way in which the system should be operated to achieve good results. It is only through accurate knowledge of the actual electrical output quantities that proper utilization of an electrical fishing system can be attained.

Nature of Flow of Electricity

According to the modern theory, which has been established by the experimental results of many of the investigators that the atom of all the matters consist of a positively charged nucleus around which negative charges rotate with high angular velocity. These individual negative charges *electrons* are indivisible and are found to be identical in all the substance. In conduction some of these electrons are free to pass from one atom to the other, when a difference of potential is impressed across the ends of the conductors. The movement of these electrons constitutes the electric current. Hence, the electricity in motion may be considered as the electric current and is called dynamic electricity.

In a non-conductors of electricity, the electrons are closely bound to the nucleus and it is difficult to remove an electron from the atom. Hence a high difference of potential is required to remove only a few electrons from the atom and the corresponding current is also extremely small.

Electron

An electron is almost a " weightless particle of electricity " possessing a negative electric charge equal to 1.58 X 10 to the power minus 19 coulomb. Electrons in motion constitute an electric current and one ampere constitutes a rate of flow of 6.28 X 10 to the power minus 18 electrons per second. It has got the characteristics of behaving both as wave and particle and it has a small mass, only 1/1840 that of a hydrogen atom and can in suitable circumstances travel almost with the speed of light. They may be symbolized by concentric spheres, in which they are free to revolve round the nucleus.

Electrical Resistance

The flow of current through a circuit not only depend on the electromotive force impressed on the circuit, but it also depends on the properties of the circuit. For example, if a copper be connected across the terminals of the battery, a current will flow through the wire. Now if the copper wire be cut and a small incandescent lamp be inserted in the circuit, the lamp filament will be heated and may become incandescent. In this case, the current in the circuit decrease with the insertion of poor conducting medium even with a constant electromotive force.

This property of an electric circuit tending to prevent the flow of current and at the same time causing electric energy to be converted into heat energy is called resistance.

Conductors and Insulators

As it is already stated that in some substance, the electrons are able to readily pass from atom to atom and is known as conductors; while in the other substance, electrons can be removed from the atom with much difficulty and are known as insulators. But all the substance offer some resistance to the flow of current and are therefore, not perfect conductors; while most of the insulators are conducting current to some extent and are not perfect insulators.

Conductors may be divided into three general classes, *metallic, electrolytic* and *gaseous*. In metallic conductors the conduction of current is due to the inter-atomic movement of the electrons within the conductor and is not accompanied by any movement of material through the conductor or by chemical action. With electrolytic conductors conduction is accompanied by a movement of material through conductor and is usually by chemical action. With gaseous conductors, conduction is due to the movement of free positive ions and free negative ions or electrons, into which the atoms of the gas become divided when it becomes ionized.

Of all the three classes of conductors, metallic conductors are best ones, silver, copper and alloys. Silver is the best conductor of all the usual metals and the copper comes to next. The other metals and their alloys have varying degree of conductivity.

The electrolytic conductors include, the solution of acids, bases and salts. Salts, oils, glass, silk, paper, cotton, ebonite, fiber, paraffin, rubber, plastic substances when dried may be considered as non-conductors or insulators.

Unit of Resistance

The ohm is the practical unit of resistance, and is defined as the resistance which will allow one ampere of current to flow if one volt is impressed across its terminals. The unit is named for Georg Simon Ohm (1787-1854) of Germany, the mathematician who about 1827 evolved the principle, now known as Ohm's law.

The resistance of a conducting body varies with the length with constant cross section and varies inversely with the cross section of the conducting body with constant length. Hence it follows that the resistance of a homogenous body of uniform cross section varies directly with its length and inversely as its cross section, the length being taken in the direction of current and the cross section perpendicular to the direction of the current.

Hence, $R = P \dfrac{L}{A}$

where, R is the resistance in ohms, L is the length in the direction of the current, A is tha area at right angles to the direction of the current and P is a constant of the material known as the *resistivity* or *specific resistance* which differs from material to material.

Electrical Properties of the Metals and Alloys

Metals	Resistivity at 20°C		Temperature Coefficient of Resistance at 20°C
	Cm Cube Micro Ohms	Cir-mil-feet Ohms	
Aluminum	2.828	17.02	0.0039
Antimony	41.7	251.0	0.0036
Bismuth	110.0	663.0	0.004
Carbon (graphite)	720-812	–	–
Copper	1.724	10.37	0.00393
Gold	2.44	14.7	0.0034
Iron (cast)	74.4-97.8	448-588	–
Lead	20.4	123	0.00387
Mercury	94.07	566	0.00072
Nickel	9.97	60.0	0.oo5
Platinum	10.96	66.0	0.003
Silver	1.628	9.796	0.0038
Tungsten	5.51	33.2	0.005
Zinc	5.75	34.6	0.0040
Alloys			
Brass	6.17	37.1	0.0015
Bronze	4.16	25	0.0020
German silver	33.0	198.5	0.0004
Manganin			
Cu-Mn-Ni	48.2	290	+0.000015

Conductance

Conductance is the reciprocal of resistance and may be defined as being that property of a circuit or of a material which tends to permit the flow of electricity. The unit of conductance is the reciprocal ohm or *mho*. Conductance is usually expressed by g

G = I/R = r A/L

Where, r is the specific conductance of a substance; A the uniform cross section and L is the length.

The conductivity of copper at 20 degree Celsius is 580000 mho-cm^{-1}

Temperature Coefficient of Resistance

The resistance of copper and other non-alloyed metals increases appreciably with temperature. It is important to know the relation between temperature and resistance as the temperature at which the electrical conductors operate varies with the current and the surrounding temperature. The temperature of the various electrical devices, such as, wires and cables, the filament of incandescent lamps and the wires in a resistance unit increases during operation.

Hence the temperature coefficient resistance at a given temperature may be defined as the change in resistance per ohm per degree Celsius change in temperature from the given temperature.

Temperature Coefficient of Copper at Various Initial Temperature

Initial Temperature	*Increase in Resistance per 1ºCelsius*
0	0.00427
5	0.00418
10	0,00409
15	0.00401
20	0.00393
25	0.00385
30	0.00378
35	0.00371
40	0.00364
45	0.00358
50	0.00352

Practical Electrical Units

The practical electrical units are defined as follows.

The unit of quantity is known as coulomb (Q). An international coulomb is the quantity of electricity which passes any section of an electric circuit in one second, when the current in the circuit is one international ampere. The coulomb is analogous to the unit quantity of water in hydraulics such as, cubic foot and the gallon.

The unit of current ia ampere (A). An ampere is equal to one coulomb per second.

The international ampere is defined as the current which under definite conditions will deposit silver at the rate of 0.001118 gram per second, when silver nitrate is electrolyzed. A current of one ampere flows when a potential difference of one volt is applied to a resistance of one ohm. It is to be kept in mind that the ampere is the rate of flow of electricity. It corresponds in hydraulics to the rate of flow of water, which is expressed as cubic feet per second or gallons per minute.

The unit of resistance is the ohm (R). The international ohm is defined as the resistance of a column of mercury of uniform cross section, having a length of 106.300 centimeters and a mass of 14.4521 grams at zero degree Celsius.

The unit of potential difference is the volt (V). The international volt is the potential difference that will produce a current of one international ampere through a resistance of one international ohm. The mechanical analogue of potential difference is pressure. The difference in hydraulic pressure between the ends of a pipe causes or tend to cause the flow of water. Similarly electric pressure or difference of potential tends to cause electricity to flow, thus producing current.

The unit of electrical power is the watt (W). The international watt is the power expended when one international ampere flows between two points having a potential difference of one international volt.

The unit of electrical energy is the joule (J). The international joule is the energy required to transfer one international coulomb between two points having a potential difference of one international volt.

Ohm's Law

Ohm's law states that for a steady current, the current in the circuit is directly proportional to the total emf (electromotive force) acting in the circuit and is inversely proportional to the total resistance of the circuit.

This law may be expressed by the following equation; if the current I is in ampere, the emf E is in volts, and the resistance R is in ohms;

$$I = E/R \text{ ampere} \tag{1}$$

That is the current in ampere in a circuit is equal to the total emf of the circuit in volts divided by the total resistance of the circuit in ohms.

By transformation, Equation (1) becomes;

$$E = IR \text{ volts}$$

That is the voltage across any part of a circuit is equal to the product of the current in amperes and the resistance in ohms, provided that the current is steady and that there are no sources of emf within this part of the circuit.

Again if the equation (1) be solved for the resistance, the result is;

$$R = E/I \text{ ohms}$$

That is the resistance of a circuit or any part of a circuit, is equal to the voltage divided by the current, provided that the current is steady and that there are no sources of emf within the part of the circuit considered.

Principle of the Electric Battery

An appreciable deflection of the voltmeter will be observed when one copper strip and one zinc strip are immerged in a dilute sulphuric acid solution instead of two copper strips. It will be necessary to connect the copper to the positive terminal and the zinc to the negative terminal of the voltmeter. This shows that so far as the external circuit is concerned, the copper is positive to the zinc.

Earlier discoveries of Galvini and Volta named the batteries or cells as galvanic batteries and voltaic cells. Solutions, such as are used with batteries are called electrolytes, or electrolytic conductors and are defined as conducting medium in which the flow of electric current is accompanied by the movement of matter.

In order to obtain difference of potential between two metal plates only the following two conditions are necessary.

1. The plates must be of different metals;
2. They must be immerged in some electrolytic solution, such as, an acid, an alkali or salt.

Electrolytic Cells

An electrolytic cell is a unit apparatus designed for carrying out an electrochemical reaction and includes a vessel, two or more electrodes, and one or more electrolytes.

An electrode is a conductor belonging to the class of metallic conductors, but not necessarily a metal through which current enters or leaves an electrolytic cell.

An anode is an electrode through which current enters conductor of the non-metalic class.

A cathode is an electrode through which current leaves any conductor of the non-metalic class.

In the last two definitions the " Conductor of the non-metalic class" is obviously the electrolyte. In an electrolytic cell delivering energy, the anode is the negative terminal or zinc and the cathode is the positive terminal or copper.

A primary cell is a cell designed to produce electric current through an electrochemical reaction which is not efficiently reversible and hence the cell, when discharged cannot be effectively recharged by an electric current.

A storage cell is an electrolytic cell for the generation of electric energy in which the cell after being discharged may be restored to a charged condition by an electric current flowing in a direction opposite to the flow of current when the cell discharges. A storage cell is frequently called a secondary cell.

Energy stored in electrolytic cells is chemical energy and electrical energy is delivered at the expense of the electrodes which either goes into the solution or is converted into lower form of chemical energy. Therefore, a primary cell or battery transforms chemical energy into electrical energy.

Batteries in Electric Fishing

The batteries used in electric fishing systems are usually sealed lead-acid storage batteries similar to those used in automotive applications. They are usually 12-volts units. The capacity is measured in ampere-hours with a wide range of sizes readily available. Deep discharge batteries are preferred since this variety is intended for the type of service which occurs in electric fishing where the battery is typically used until it can no longer provide the necessary power to operate the system (in contrast to automotive use where the battery is kept at a nearly full state of charge). Deep discharge batteries can be fully cycled many times (over 500) without damage.

Batteries have an *external characteristic* (a plot of voltage vs current) like that shown in Figure 2.1. In common with all types of electric power supplies, the available voltage decreases as current is delivered to a load. As the battery discharges, the available voltage at any specific current decreases until the voltage is no longer sufficient to adequately power the load. This is illustrated in Figure 2.1 by the curve labeled "discharged". This means that the available voltage gradually declines as the battery discharges giving rise to a gradual decline in the performance of the electric fishing system (some power conditioners have voltage regulating circuit to compensate for this effect and maintain performance until the battery reaches a lower bound where operation ceases).

The actual ampere-hour capacity of a battery depends on a number of factors such as temperature, rate of discharge, age etc. The effect of the rate of discharge is illustrated in Figure 2.1 where it can be seen that high discharge rates lead to reduced

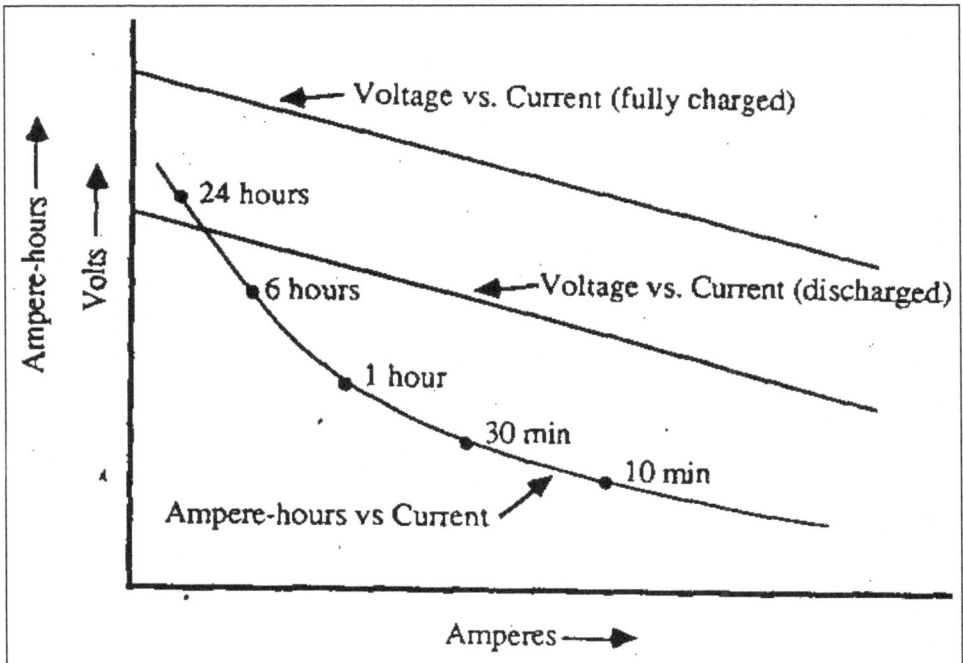

Figure 2.1: External characteristic and capacity of batteries.

ampere-hour ratings. These characteristics vary greatly depending on battery design and quality and hence on cost.

The Generator

An electric generator is a machine which transforms mechanical power into electrical power. This is accomplished by means of an armature carrying conductors on its surface, which act in conjunction with a magnetic field. Electromotive force is induced by the relative motion of the armature conductors and the magnetic field, and when the current is delivered to an external circuit, electric power is supplied by the armature.

In the D.C. generator the field is usually stationary and the armature rotates. In most types of A.C. generator, the armature is stationary and the field rotated. Either the armature or the field is driven by mechanical power applied to its shaft.

Electromotive Force Generated by a Moving Coil

The action of the generator is based on this principle. The flux linking the armature coils is varied by the relative motion of armature and field. A coil of single turn is shown in the Figure 2.2.

The coil rotates in an anticlockwise direction at a uniform speed in an uniform magnetic field. As the coil assumes successive positions, the emf induced in it changes the position. In position 1, the emf generated is zero, for in this position, neither active conductor is cutting magnetic lines, but is moving parallel to these lines. When the coil reaches position 2 its conductors are cutting across the line obliquely and the induced emf has a value indicated at 2 in Figure 2.2(b). When the coil reaches the position 3, the conductors are cutting the lines perpendicularly and are cutting therefore, at the maximum possible rate. Hence the induced emf is the maximum when the coil is in this position. At the position 4 again the induced emf is less, due to lesser rate of cutting. At position 5 no lines are being cut as in case of 1 and the induced emf is zero. In position 6, the direction of the induced emf in the conductors will be reversed, as each conductor is under a pole of opposite sign to that for positions 1 to 5. The induced emf increases to a negative maximum at 7 and then decreases until the coil again reaches position 1. The region corresponding to positions 1 and 5 in which no emf is induced in the coil by rotation is called neutral zone. Since the induced emf reverses in direction cyclically, it is an alternating emf and an emf varying in a manner shown in Figure 2.2(b).

This alternating emf may be impressed on an external circuit by means of two slip rings (Figure 2.3). Each ring is continuous and is insulated from the other ring and from the shaft. A metal or carbon brush rests on each ring and conducts the current from coil to the external circuit.

In order to receive direct current, one whose direction is always the same, such slip rings can not be used. The coil current must be alternating, since the emf produced by it is alternating as shown in Figure 2.2(b). The current must be rectified before it enters the external circuit. This rectification can be accomplished by using a split ring (Figure 2.4(a). Instead of two rings as in Figure 2.3, one ring is used. This is split by cut

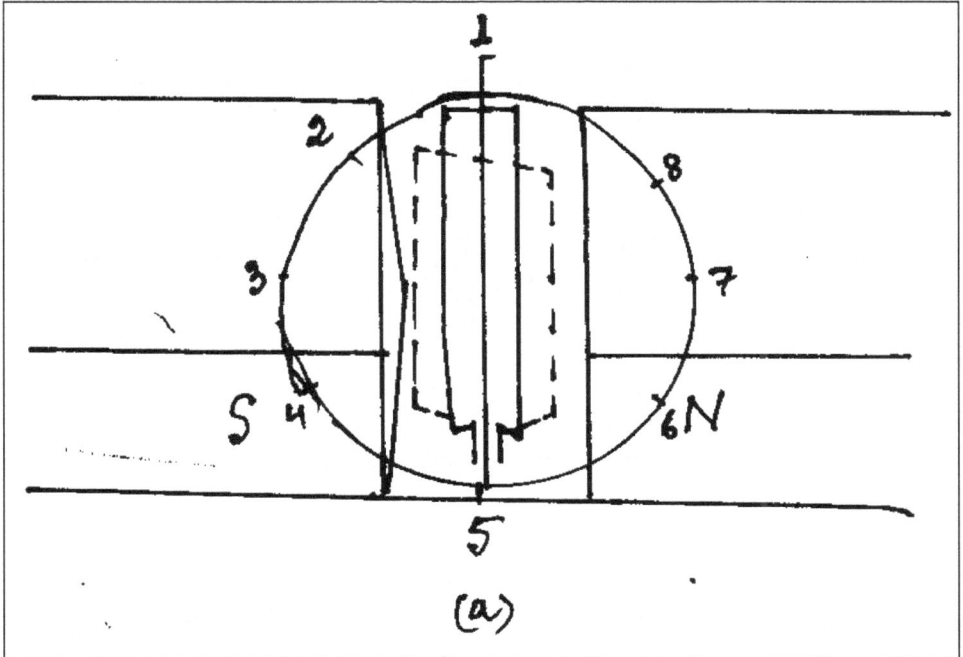

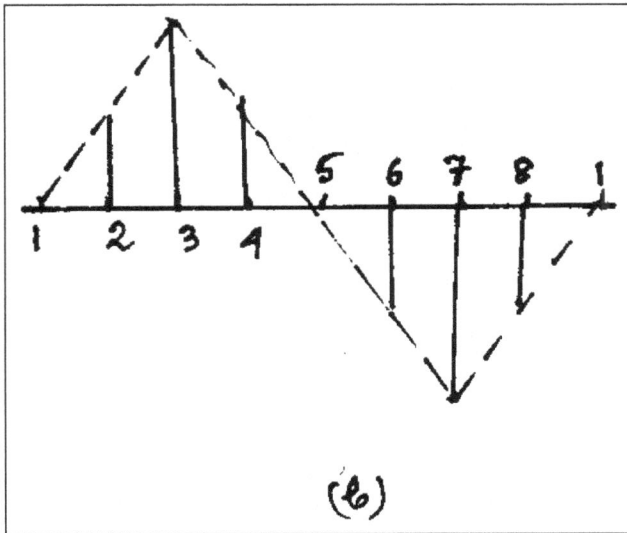

Figure 2.2(a-b): Electromotive force induced in coil rotating at constant speed in uniform magnetic field.

at two points diametrically opposite, and the ends of the coil are connected to the sections or segments so produced.

A consideration of Figure 2.4(a) shows that as the direction of the current in the coil reverses, the connections to the external circuit are simultaneously reversed.

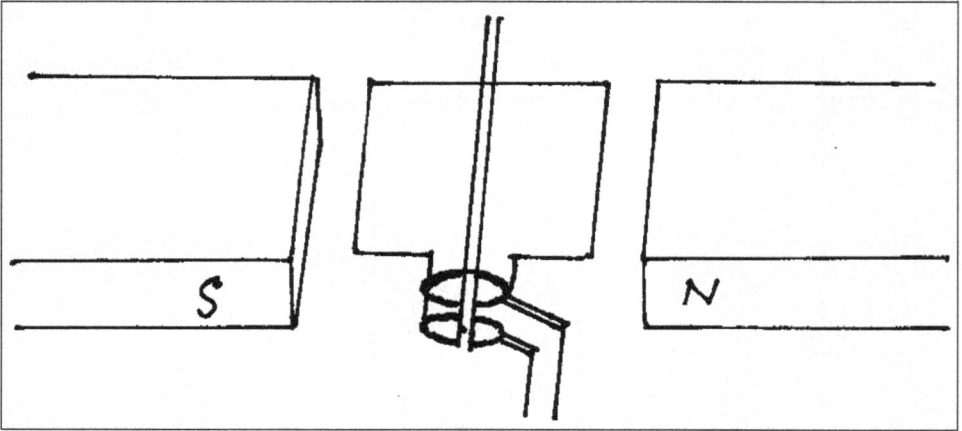

Figure 2.3: Current taken from rotating coil by means of slip rings.

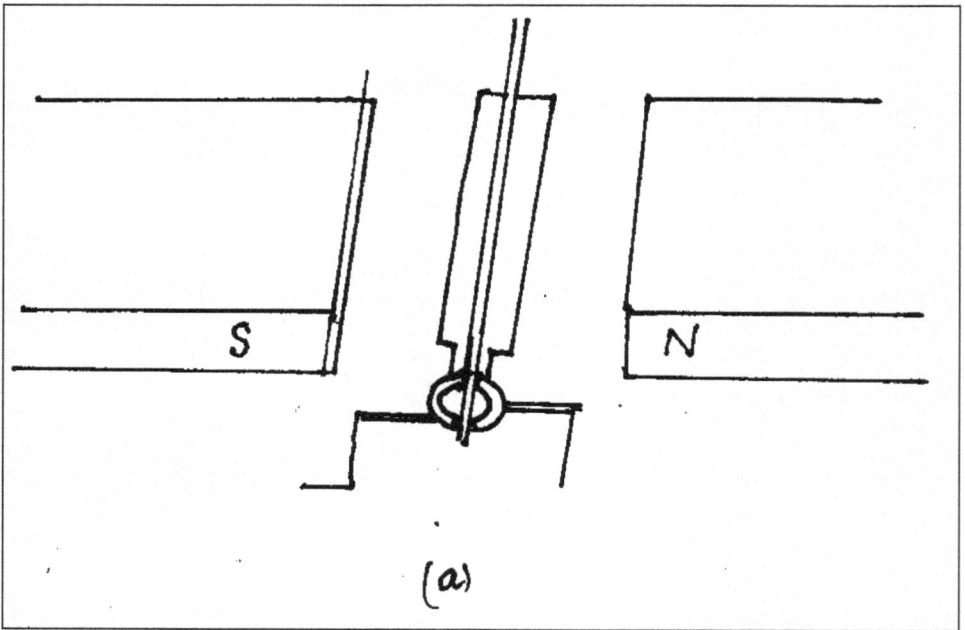

(a)

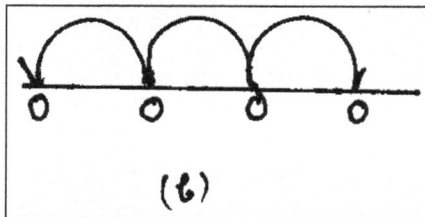

(b)

Figure 2.4(a-b): Rectifying with split ring.

Therefore. The direction of current in the external circuit is not changed. The brushes pass over the cuts in the ring when the coil is perpendicular to the magnetic field, or in so called neutral zone and no emf is produced as at points 1 and 5 of Figures 2.2(b). comparing 2.2(b) with Figure 2.4(b), it is seen that the negative half of the wave has been reversed and made positive.

D.C. Generators

Unlike batteries, generators can be used without a power conditioner to directly provide the excitation for electric fishing. This is very common in DC electric fishing in streams and lakes where a DC generator is often directly coupled to the electrode system without any intermediate power conditioner or electrical instrumentation. In these systems the DC generator is often used to supply lights or other peripheral loads. This practice should not be permitted for safety reasons; a separate, low voltage power source should always be provided for these auxiliary loads.

A true DC generator is an electric machine which employs a commutator and brushes to mechanically convert the AC power produced inside the machine to DC power at the terminals. There are also machines which use an electronic rectifier to convert the AC power from the machine to DC. These latter machines produce an output voltage which contains a large ripple component and are not well suited to DC electric fishing (they produce more nearly the type of response associated with AC). An electric filter can be added between the rectifier and the electrodes to remove the ripple if DC electric fishing is to be attempted with such machine. The true DC generator has only a very small ripple component.

Figure 2.5 shows the external characteristics (voltage v. current) for a typical true DC machine.

In any electric generator the output voltage drops as the generator is placed under load. First there is an internal voltage drop which increases as the current increases (as in a battery), and second, placing a load on the engine causes the engine speed to drop. The amount of voltage reduction (or voltage regulation, usually expressed in per cent) depends on the type of generator, the quality of the speed governor on the engine, the condition of the engine, and to some extent on the size of the generator. The name plate *rated voltage* of any generator is always for full nameplate *rated current* and, hence, the open circuit or no load voltage will always be higher than the rated voltage as illustrated in Figure 2.5. For a true DC generator the voltage regulation is of the order of 5 to 15 per cent. It is typically much higher in an AC generator with an electronic rectifier.

The effect of speed reduction on the external characteristic is also shown in Figure 2.5. In a system where the DC generator directly feeds the electrodes, speed variations provide some control over the output voltage but in most cases the range is quite limited. In some machines the field current, which sets up the magnetic field to generate the output voltage, can be varied to control the output voltage level. This is however, not common on small commercially available generators.

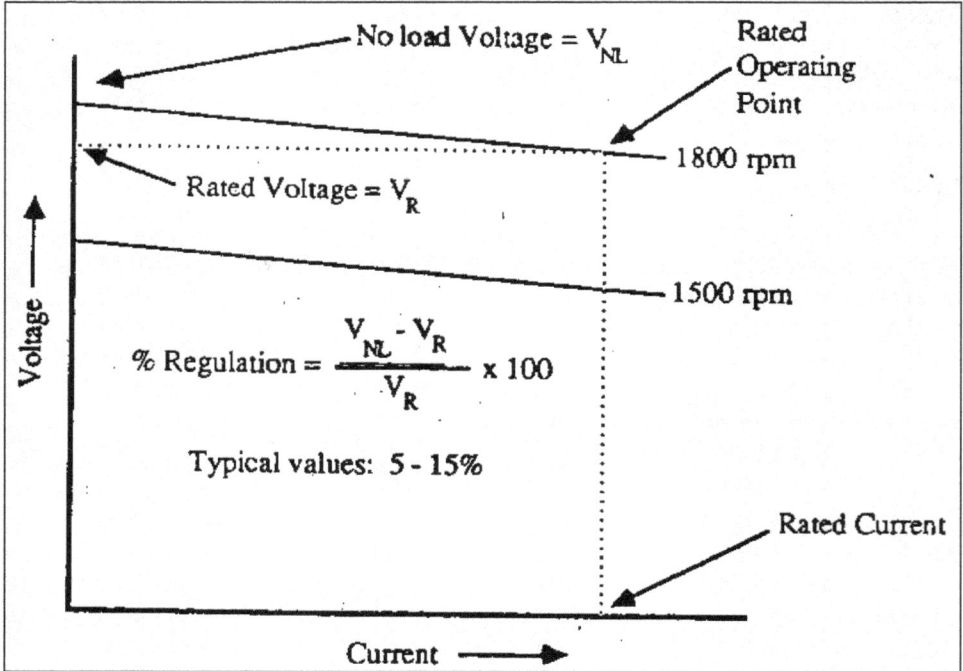

Figure 2.5: Typical external characteristics of DC generators.

AC Generators

Alternating current generators do not utilize a commutator and brush assembly and are therefore more rugged and reliable machines. The output waveform is typically sinusoidal with the most common frequencies being 50, 60, 180 and 400 Hz (cycles/second). There are two basic types; single phase and three phase. A single-phase generator has a single winding and a single pair of output terminals which supplies the entire output power of the machine. Three-phase generators have three sets of windings usually with one end of each winding connected to each other to form a common point or neutral as shown in Figure 2.6. Typically the neutral is not accessible and only the three free ends of the windings are available at the machine terminals. In some machines the neutral is connected to the metal frame of the generator; for electric fishing this "frame ground connection" must be removed for safety reasons.

The output voltages of a three-phase generator are equal in amplitude and time displaced from one another by one-third of a period to form what is called a balanced three-phase set (Figure 2.6). This arrangement has significant advantages from the point of view of internal generator performance. The advantages include better utilization of materials (three-phase machines are about 40 per cent smaller for the same power compared to single-phase machines), higher efficiency and a smoother, more vibration-free operation. They also provide a better form of excitation for most electronic power conditioners which usually convert the AC power to DC first. The

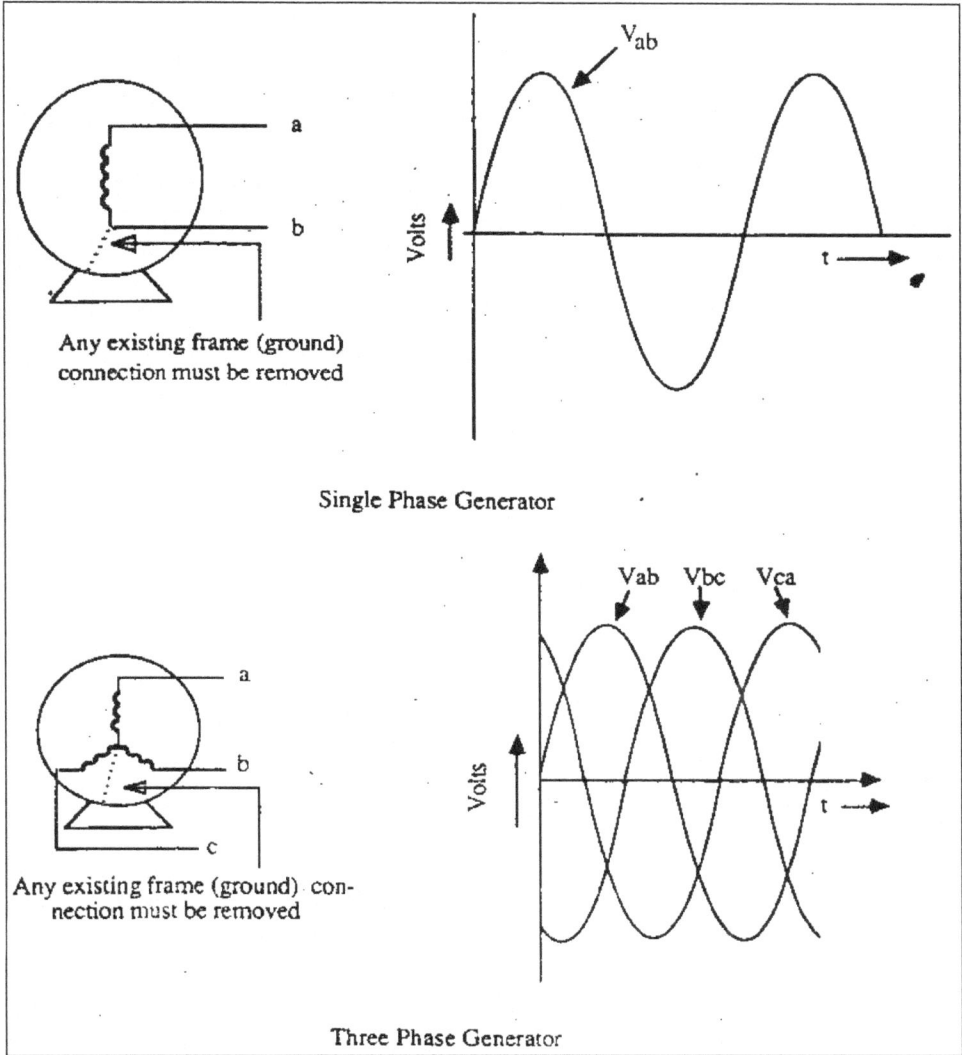

Figure 2.6: Circuit diagrams and waveforms of single-phase and three-phase generators.

three-phase supply results in much less ripple in the DC power and hence less filtering is required.

The external characteristic of an AC generator is similar to that of a DC machine except that the per cent drop in voltage (voltage regulation) is usually somewhat larger. Figure 2.7 illustrates a set of typical characteristics for two different speeds of the engine. In the AC machine the frequency as well as the voltage changes with engine speed. As in the DC machine the voltage of some AC generators can be changed by changing the field current; in this case the frequency is not affected. This can cause problems in a transformer fed from the generator if the voltage is raised much beyond

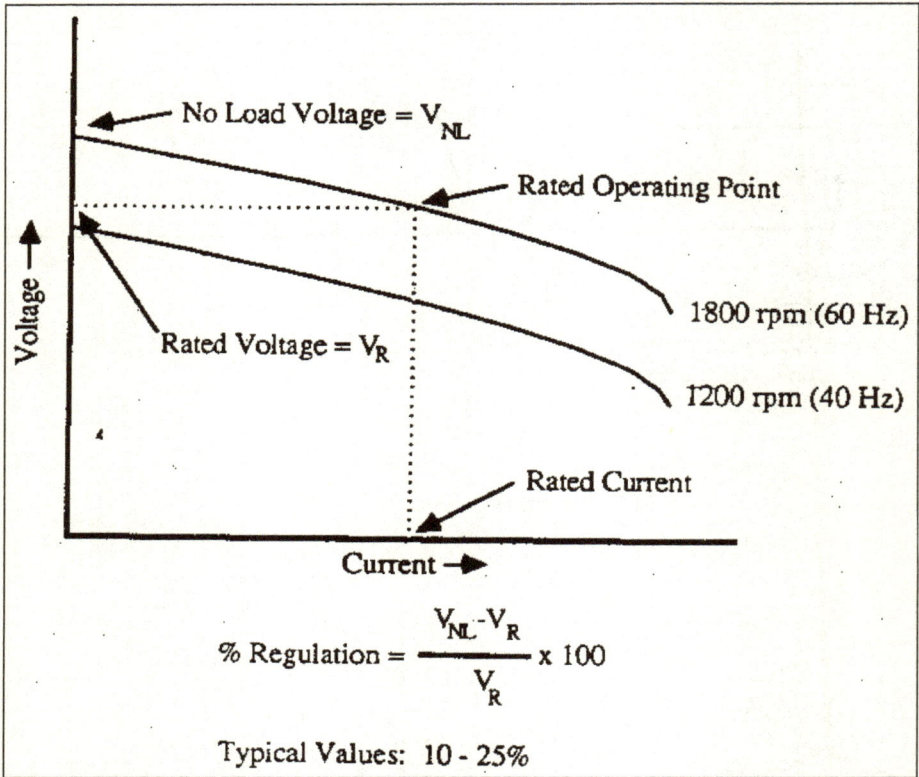

Figure 2.7: Typical external characteristics of AC generators.

the rated voltage of the transformer. There is no problem when both voltage and frequency are increased together (as when generator speed is increased) because transformers are only sensitive to the ratio of voltage to frequency (often stated as V/ hz).

A three-phase generator is the preferred machine for most electric fishing applications primarily because it is smaller and lighter (for the same power) and better suited as a supply for most power conditioners. In all cases it is important to try to attain roughly equal currents in the three phases since unbalanced operation can create excessive heating and vibration in the machine. Small amounts of unbalance are acceptable, but the generator performance is always best under balanced conditions. A transformer-rectifier system provides a voltage-controlled DC source without the need for large filters as would be required with a single-phase generator.

Compared to a DC generator, the ease of controlling voltage with transformers to match the generator to water conductivity conditions is a major advantage of an AC generator. Since higher frequencies offer a weight advantage in the generator and in designing suitable transformers, the highest commercially available frequency is preferred.

Transformers

The static transformer is a device for transferring electrical energy from one alternating current to another without a change in frequency. This transformer is usually, but not always accompanied by a change of voltage. A transformer may receive energy at one voltage and deliver it a higher voltage in which case it is called step-up transformer. A transformer energy at one voltage and deliver it at lower voltage in which case it is called step down transformer. The transformer which receive one voltage and deliver it at the same voltage, it is called one to one transformer.

The transformer is based on the principle that energy may be efficiently transferred by induction from one set of coils to another by means of varying magnetic flux, provided that both sets of coils are on a common magnetic circuit.

Electromotive forces are induced by a change in flux linkages. In the generator, the flux is substantially constant in magnitude. The flux linking the armature coils is changed by the relative mechanical motion of flux and coils.In the transformer, the coils and magnetic circuit are all stationary with respect to one another. The emfs are induced by the change in the magnitude of the flux with time.

The use of transformers to control the output voltage of AC generators, and thus provide a proper match between electrode resistance and the generator characteristic, is extremely valuable. This is particularly true when it is necessary to operate over a wide range of water conductivity.

As a general guide to the voltage range needed for various water conductivities, voltage levels found useful for AC electric fishing are presented in the table below. Since the voltage level required depends greatly on the size and type of electric fishing system, the data in this table should be considered as a general guideline for electric fishing boats similar in size and type to those described in Novotny and Preigal (1974). The power values given in the table indicate the levels employed in these boats. Except for very low conductivities, the power was limited by equipment ratings and not by electric fishing considerations.

Voltage Ranges for AC Electrofishing

Conductivity Micro-Siemen/cm	Voltage (V)	Power (kW)
10-20	460*	2.0!
20-40	460*	3.5!
40-60	460-390	4.5*
60-100	390-320	4.5*
100-160	320-230	4.5*
160-350	230-160	4.5*
350-700	160-100	4.5*

*: Limit imposed by equipment rating; !: Limit imposed by maximum electrode size.

Transformers consist of coils of wire wrapped around an iron core. In fixed ratio or tap changing transformers there are no moving parts (except for switches) or electronic devices so the reliability is very high. Continuously variable ratio transformers have a sliding brush arrangement which is less reliable in the adverse environment to which electric fishing systems are subjected. Tap changing transformers pose no real limitation since the main engine generator can be controlled over a speed range which allows intermediate voltages when necessary. In addition to being more reliable, this arrangement has a considerable weight advantage over a continuously variable transformer since the windings can be designed to be operated at nearly maximum utilization on all range settings.

While a transformer is often thought of as a device which transforms the voltage level, it is important to recognize that it also transforms the current. The simplest way to think about how a transformer operates is to remember that the volt-amperes (product of voltage times current) is the same on both sides of the transformer. (This follows immediately from conservation of energy; a transformer is not an energy source). Thus if a transformer is used to double the voltage the current on the secondary side will be one-half that on the primary side. This current transformation property is extremely important in electric fishing in very high conductivity water since it permits supplying the electrode system with much larger currents than can be supplied directly from the AC generator. The use of step-down (voltage) transformers in these very high conductivity situations has not been widely exploited and should be considered whenever such problems are encountered.

A two-winding transformer (often called an isolation transformer) rated for the same power as the generator it is connected to, will be roughly comparable in size and weight to the generator (not including the engine). Since the isolation feature is not needed in most electric fishing systems, an autotransformer arrangement can be used to reduce the physical size and cost of the unit. An autotransformer is simply a transformer connection in which the primary and secondary windings are interconnected electrically rather than being totally separated electrically. Since some of the energy is directly conducted to the output, the transformer core and coils can be smaller. The reduction in size depends on the amount change in voltage and becomes less significant as the voltage ratio becomes large. For a two to one voltage change an autotransformer is one-half the size and one-half the cost of a two winding transformer (Figure 2.8).

Transformer size and cost also depend strongly on the electrical frequency. Higher frequencies result in a substantial size reduction and are therefore preferred for portable systems.

Static Electricity

So far the electricity in motion or dynamic electricity has been considered. It will not be beyond the scope of the book to have a fundamental idea about the static electricity or electricity at rest.

As discussed earlier that in dynamic electricity electrons move from atom to atom in a conducting medium. With the static electricity, the effect of the electrons is

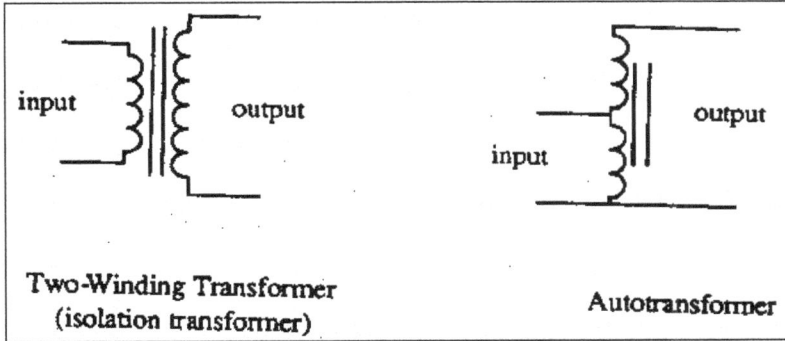

Figure 2.8: Transformer types, isolation transformer and autotransformer.

as if they are at rest although there may have been an initial movement of electrons with respect to the atom.

For example, if the insulated parallel conducting plates A and B (Figure 2.9) are connected to the positive and negative terminals of the battery, the positive of the applied potential will attract or withdraw, some of the free electrons from electrode A and the negative terminal will repel electrons to electrode B. Hence the battery or the source of potential has withdrawn electrons from electrode A and transferred them to electrode B. Thus plate A has become positively charged since negative charges have been drawn from it, and the plate B has become negatively charged. Under these conditions the charge on the plates is called static electricity.

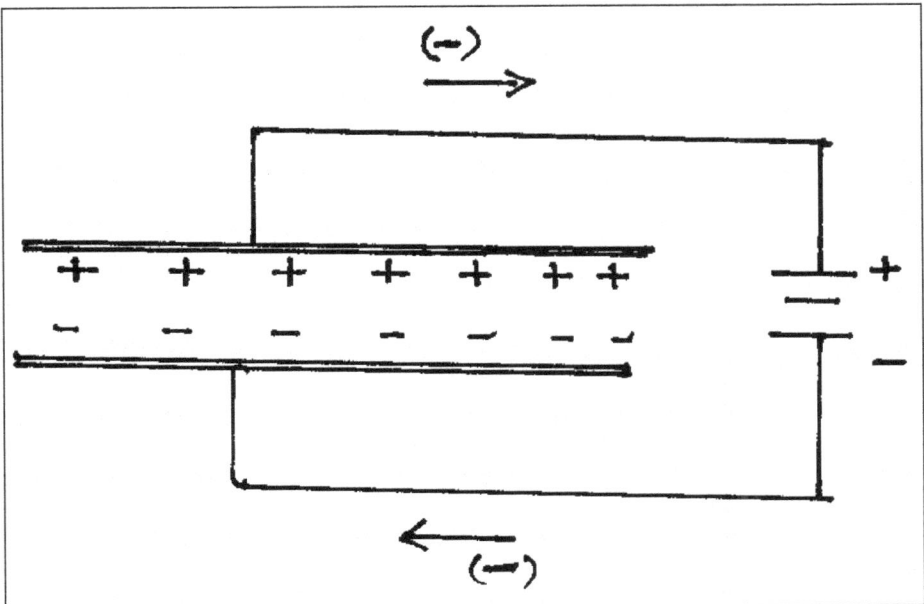

Figure 2.9: Transfer of electrons in capacitor plates.

So it appears from the above that dynamic and static electricity are identical in their ultimate nature. With dynamic electricity, there is a movement of electrons between the adjacent atoms of the conductor. With static electricity, free electrons have been displaced from the positive to the negative plate and are maintained in this condition by the electric field. This displacement has the effect of producing a single negative charge on the negative plate and a single positive charge on the positive plate. This displacement of charges at the positive and negative plates is frequently associated with high voltage.

Capacitor

Two conductors when separated by a dielectric constitute a capacitor. When a potential difference is applied between the the plates of a capacitor, electricity is stored in the capacitor, a positive charge being on one plate or set of plates, and an equal negative charge on the other plate or set of plates. This property of a capacitor to store electricity is called capacitance The relation between the voltage and the charge in a capacitor is expressed by the equation;

$$Q = CE$$

That is the charge in a capacitor is equal to the voltage multiplied by a constant C. This constant C is the capacitance of the capacitor. The practical unit of capacitance is the *farad*. If C is in farads and E in volts, Q is in coulombs.

The farad is too large a unit of capacitance for general use. The capacitance of the earth as an isolated sphere is less than one thousandth of a farad. The *microfarad* (μf), equal to one millionth of a farad is therefore used as the unit of capacitance.

Current Conduction in Liquid

Pure water is a poor conductor of electricity. In fact, it may be considered as practically insulator. If however, even a very small amount of acid, alkali or salt be added to water, the solution becomes good conductor. Current in the solution dissociates molecules of substance or the water and is known as electrolytic dissociation. In solution, the molecules of acids, alkalies and salts dissociate into positive and negative ions. According to electron theory, negative ions has excess of electrons and positive ions has deficiency of electrons. When no potential difference is applied between electrodes, the ions drift about in the solution. When a potential difference is applied to the electrodes in the solution, the positive ions migrate to the negative electrode, or cathode, and accordingly are called cations. The negative ions go to the anode or positive electrode and accordingly called anions.

The positive ions give up their charge to the cathode and the negative ions give up their charge to the anode, thus constituting the current. Hence the conduction of current through an electrolyte is a convection effect, the charges being carried to the electrodes by the ions. Thus the electrolytic conduction differs from the ordinary metallic conduction in that it involves a transfer of matter and is accompanied by chemical change.

Chapter 3
Electronic Devices

Rectifiers and Filters

A rectifier is an electronic device for converting AC electric power to DC. There are a variety of types in general use, but primarily the *full wave* types are of use in electrical fishing systems. This is because in DC electric fishing a very smooth ripple-free DC supply provides anodic protection and reduces the susceptibility to electrotetanus caused by the ripple. A simple half-wave rectifier (Figure 3.1a-b) is sometimes used to provide a simple form of pulsed DC (sine-wave pulses) but such systems are inflexible and of very limited practical value.

The waveforms resulting from single-phase and three-phase rectifiers are shown in Figure 3.1a-b. The superiority of the full wave rectifier over a half-wave circuit is quite clear in terms of both greater average voltage and reduced ripple. A significant additional improvement is obtained in the three-phase rectifier where there are no points of zero output voltage and clearly relatively little ripple. In most cases the rectifier will be followed by an electrical filter to reduce the ripple. This is especially important if the DC output is to be used directly for DC electric fishing. Totally unacceptable DC electric fishing will result if a single phase rectifier is used without a good filter. In some cases a three-phase rectifier can be used without a filter but better results are usually obtained by including a filter. Figure 3.1a-b illustrates the waveforms obtained when a filter is used. The filter is simply a capacitor with a parallel connected 'bleeder' resistor to drain off the charge on the capacitor when the system is turned off (the load can serve as a bleeder if it can not be disconnected before the power is turned off). The filter for a three phase rectifier is smaller and less costly than for a single phase rectifier since there is much less ripple to be removed in the three phase system. An AC generator with a well-filtered rectified output will provide the same DC electric fishing performance as a DC generator with the same output voltage and power rating.

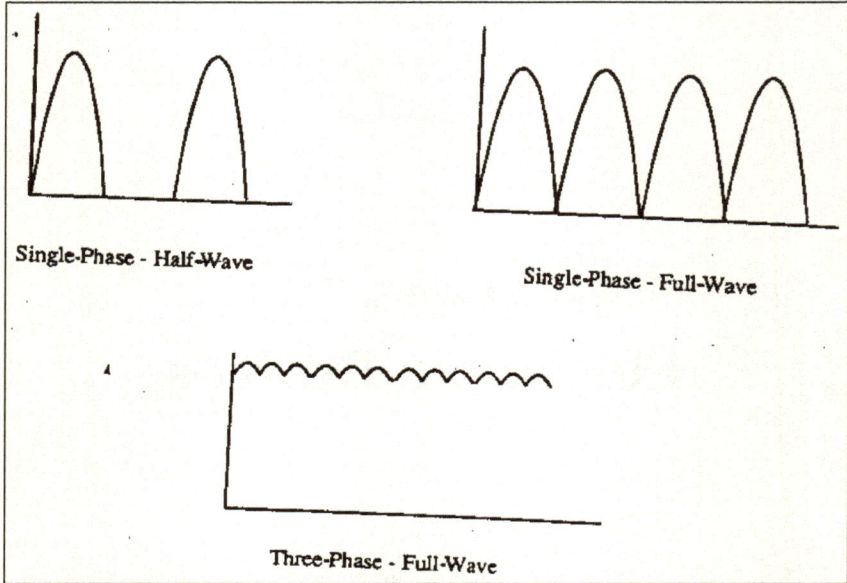

Single-Phase - Half-Wave

Single-Phase - Full-Wave

Three-Phase - Full-Wave

Figure 3.1a: Rectifier output waveforms.

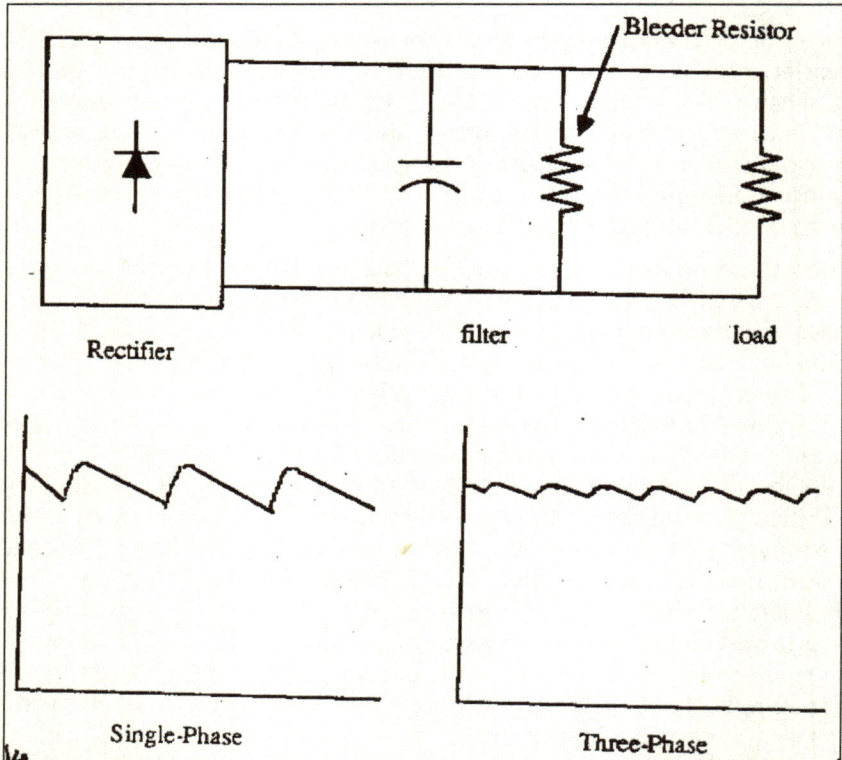

Bleeder Resistor

Rectifier

filter

load

Single-Phase

Three-Phase

Figure 3.1b: Circuit schematic and output waveforms of rectifiers with filters.

Pulsed DC Systems

Power conditioners which supply pulsed DC output belong to a class of power electronic convertors known as choppers. In electric fishing three basic types of units are in common use and are distinguished by the three different output waveforms which are produced. Figure 3.2 illustrates these three different waveforms and gives the relations for the average rms (root mean square of effective) value for each wave.

Quarter-Sine Wave Chopper

The quarter-sine wave chopper is the simplest and least expensive form of pulsed DC power conditioner. In its simplest form it consists of an uni-directional electronic switch (thyristor) in series with the AC supply (single-phase) feeding the unit. This switch is controlled to turn on only when the AC sine wave is close to its peak value. The device then continues to conduct until the current reaches zero and then returns to the off state. This operation is repeated each cycle of the input sine wave. The resulting output waveform is the quarter-sine wave (Figure 3.2).

There is some evidence to indicate that the fast rise and slow decay of the quarter-sine wave pulse is advantageous for electric fishing. While there may be some advantages in terms of fish response, the primary reason for the wide use of the quarter-sine wave chopper is its simplicity and low cost. It is, however, very limited in that it can only produce a single pulse rate equal to the input sine wave frequency and a single duty cycle of 25 per cent. A somewhat more complex arrangement using the equivalent of a full-wave rectifier plus the series switch can operate at two pulse rates(equal to the input frequency or twice the input frequency). A modification to allow the switch to close at an adjustable point in the input wave is also available and allows control of the amplitude of the voltage by controlling the length of the pulse. However, while the quarter-sine wave chopper is a reliable, low-cost pulsed DC supply it does not offer the flexibility associated with other types of choppers.

DC Chopper (Rectangular Wave Pulser)

The conventional DC chopper consists of an uni-directional electronic switch in series with a DC supply. In its most common electric fishing form it is fed from an AC supply (single or three-phase) and contains a full-wave rectifier and filter followed by the chopper. The output wave form is shown in Figure 3.2. The electronic switching element in this case must be capable of both turn-on and turn-off control as opposed to only turn-on control as in the quarter-sine wave chopper. The commonly used devices are transistors or gate turn-off thyristors. Because of the larger number of electronic components, the requirement for turn-off capability and more complex control circuits, the DC chopper is substantially more costly than the quarter-sine wave chopper. However, the DC chopper is much more flexible in that a wide range of pulse frequencies and pulse durations can be made available. With today's electronic components any pulse rate or pulse duration likely to be useful in electric fishing can be provided without difficulty (including pure DC).

Most modern rectangular-wave pulsers provide a wide range of pulse rates in the range from DC up to 200 to 500 pulses per second (pps) and duty cycles from 10

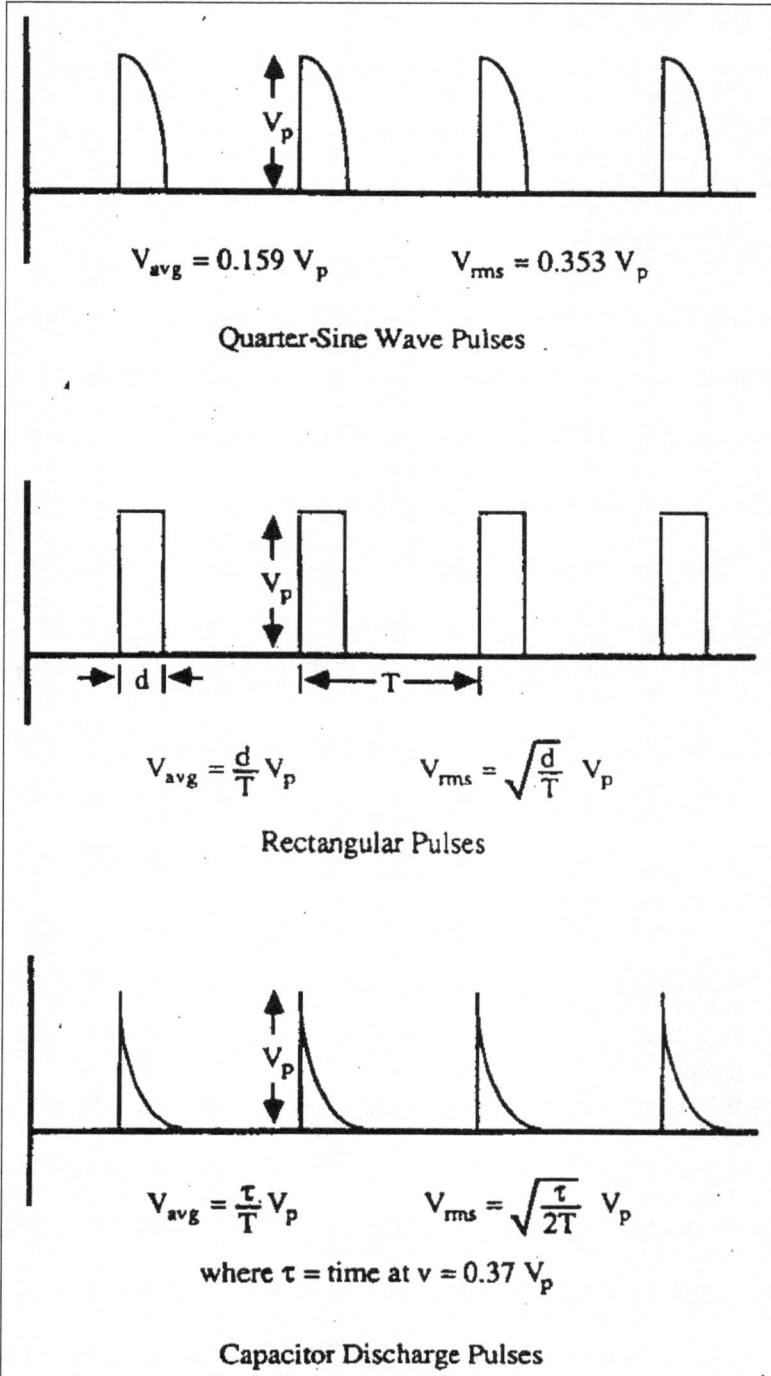

$$V_{avg} = 0.159\ V_p \qquad\qquad V_{rms} = 0.353\ V_p$$

Quarter-Sine Wave Pulses

$$V_{avg} = \frac{d}{T}\ V_p \qquad\qquad V_{rms} = \sqrt{\frac{d}{T}}\ V_p$$

Rectangular Pulses

$$V_{avg} = \frac{\tau}{T}\ V_p \qquad\qquad V_{rms} = \sqrt{\frac{\tau}{2T}}\ V_p$$

where τ = time at $v = 0.37\ V_p$

Capacitor Discharge Pulses

Figure 3.2: Output waveforms of pulsed DC power conditioners.

per cent to 50 per cent. The most commonly used ranges are from 5 to 50 pps with duty cycles from 10 to 25 per cent.

In its basic form the DC chopper does not provide control of the pulse voltage and thus voltage control must be provided by other means (the average voltage can be controlled by varying the duty cycle but this is not sufficient for electric fishing). The most common method of pulse voltage control is to vary the input AC voltage to the chopper rectifier by means of a transformer. In smaller units intended for battery powered systems, the input from the battery is *inverted* to AC first. A transformer is then used to obtain the required voltage and the output is fed to the rectifier. Higher power systems of this type will become available in the near future (fed from a generator) and will offer the advantage of much lighter weight as a result of using higher frequency transformers along with continuously variable voltage instead of discrete voltage steps.

DC choppers are very flexible in terms of controlling the pulse-on and off times. This means it is possible to produce pulse trains with special forms such as, a string of pulses at one rate followed by a second string at a different rate (or duty cycle). Some experimental work on special pulse trains has been carried out but the advantages have not been resulted in wide use of such systems as yet.

Capacitor Discharge Choppers

The third type of pulsed DC power converter is a system which utilizes a capacitor across the output terminals to produce a fast rise, slow decay output pulse as shown in Figure 3.2a. Electric energy is transferred to the capacitor in a controlled fashion by a series switching device like in a DC chopper. The stored energy is then discharged into the output (electrode system) at a rate controlled by the capacitor size and the electrode system resistance. When the electrode resistance is small (current large) the pulse decays rapidly. Most systems of this type are low power systems, usually battery supplied. The advantages are reduced power consumption for a given peak voltage and enhanced fish response (electrotaxis).

The improvement in fish response is difficult to assess since quantitative data are difficult to obtain. There does seem to be evidence that for an equal power output the capacitor discharge pulse is more efficient and thus has an advantage in battery powered equipment. Unfortunately, this effectiveness is realized in increased injury to the fish, particularly with respect to broken vertebrae. At the low power levels used in such equipment the circuitry is easily implemented and is becoming more widely used. For the higher-power levels used in generator supplied systems the electronic circuitry is more difficult to implement and more costly and hence not widely used at present. Development of high-power systems can be expected and may offer the advantages of reduced weight and power consumption. Capacitor discharge systems are less flexible with regard to pulse-on and off time than DC choppers but substantial control potential does exist. Development of special pulse trains can also be expected in these systems.

Electrical Instrument and Measurements

Moving Coil Instrument

It is an electrical measuring instrument in which a coil carrying a current is acted upon and caused to move by a permanent magnet. The same type of movement serves for a milliamperemeter, ammeter or voltmeter. The actual current in the moving coil is in all cases will be small, often one milliampere. To make a voltmeter, the current is limited by means of a large series-resistance, whereas for ammeter and milliamperemeter, a shunt is provided internally or externally. In the latercase the leads between the instrument and the shunt form part of the measuring circuit so that readings may not be correct if the wrong leads are used. Moving coil instruments work on DC only. They are characterized by the uniformity of the scale divisions.

Moving Iron Instrument

This electrical measuring instrument in which the principle of action is based either upon the repulsion of two similarly magnetized pieces of soft iron, or upon the attraction of a single piece by a coil. The force depends upon currents required so that the instrument will operate on or DC and the scales cramped at the lower end. In practice the moving irons are often shaped in such a way as to give a scale which is nearly uniform.

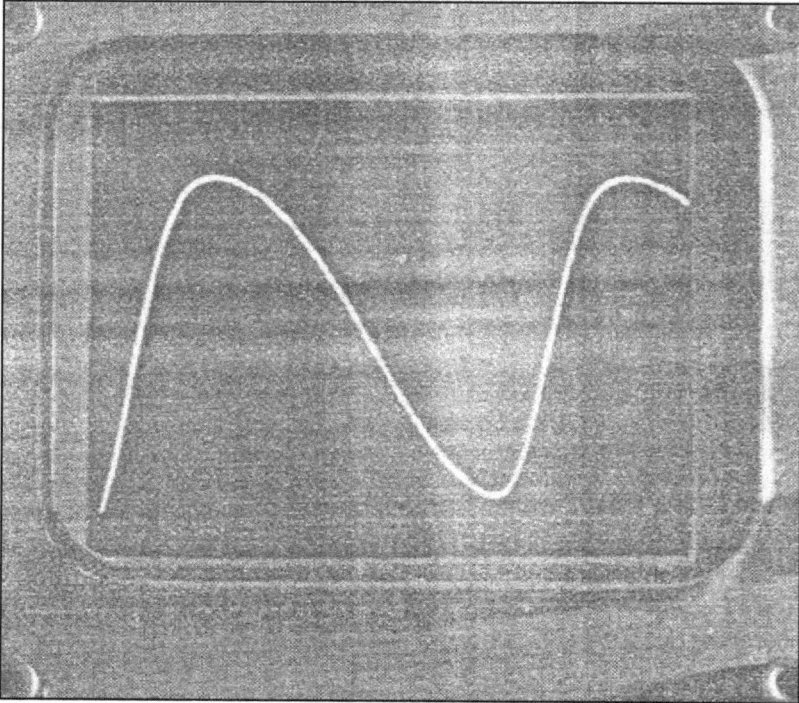

Figure 3.2a: Curve form of electrical discharge of electric eel recorded in cathode ray oscillograph.

The current required for full scale deflection depends on the number of turns in the coil. The instrument may be used as a voltmeter by putting it in series with high resistance.

Ohmmeter

It is an electrical instrument for the measurement of resistance by direct reading. It usually contains either a battery or a generator so that no separate supply of electric current is necessary One type consists essentially of current measuring device scaled in terms of resistance, while in another form the deflection depends on the ratio of voltage to current. The latter is a true ohmmeter.

Cathode-Ray Oscillograph

It is an electrical instrument by which voltages or electric currents are made to form a trace of light on a screen. The dimensions of the trace may be used as a measure of the applied voltage or current, the shape or pattern indicating the frequency and wave form. The actual indicator is a cathode ray tube, which consists of an evacuated glass vessel in which the beam of electrons emitted from the heated cathode is made to impinge on a fluorescent screen. The impact of electrons which in practice are focused to a point on the screen, causing the coating material to glow.

The beam may be deflected electrostatically or electromagnetically in a horizontal or vertical direction, in which case the visible trace would be a simple line.

In electrophysiology, the oscillograph is used to determine the wave form, and the frequencies of impulse current along with the time of impulse and pause.

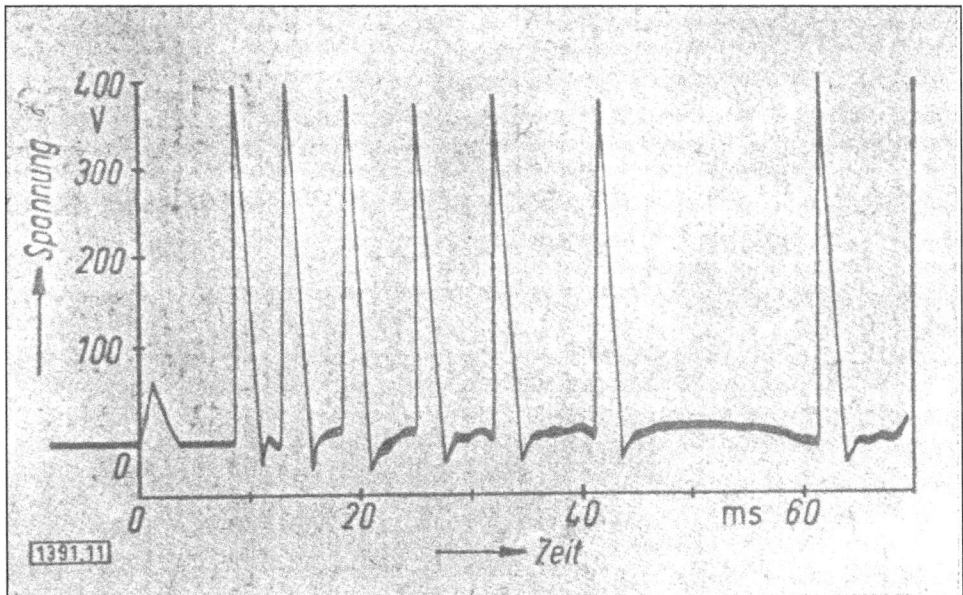

Figure 3.2b: Oscillograph tracing of discharge from electric eel.

Figure 3.2c: Laboratory pulse generator consisting of (left to right) oscilloscope for monitoring output waveform, low volage electronic pulse generator, pulse width control box with circuit for interrupting pulse trains, three power supplies, and the high-volage, high-power shocker.

Figure 3.2d: *Puntius ticto* perpendicular to bottom plane with expanded fins and bending of tail during transverse oscillotaxis.

Figure 3.2e: *Labeo rohita*–Narcotized near positive electrode.

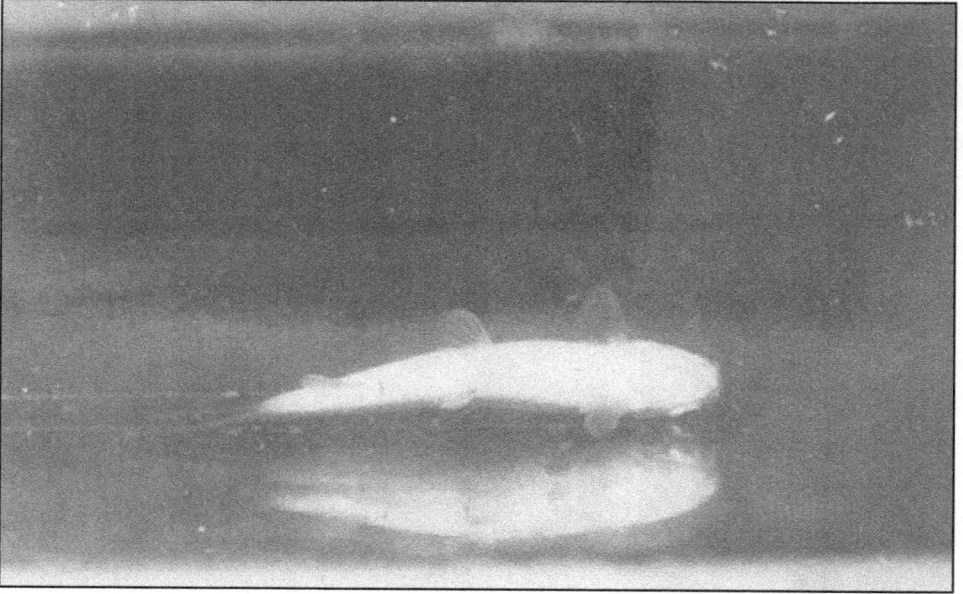

Figure 3.2f: *Cirrhinus mrigala*–Narcotized near positive electrode.

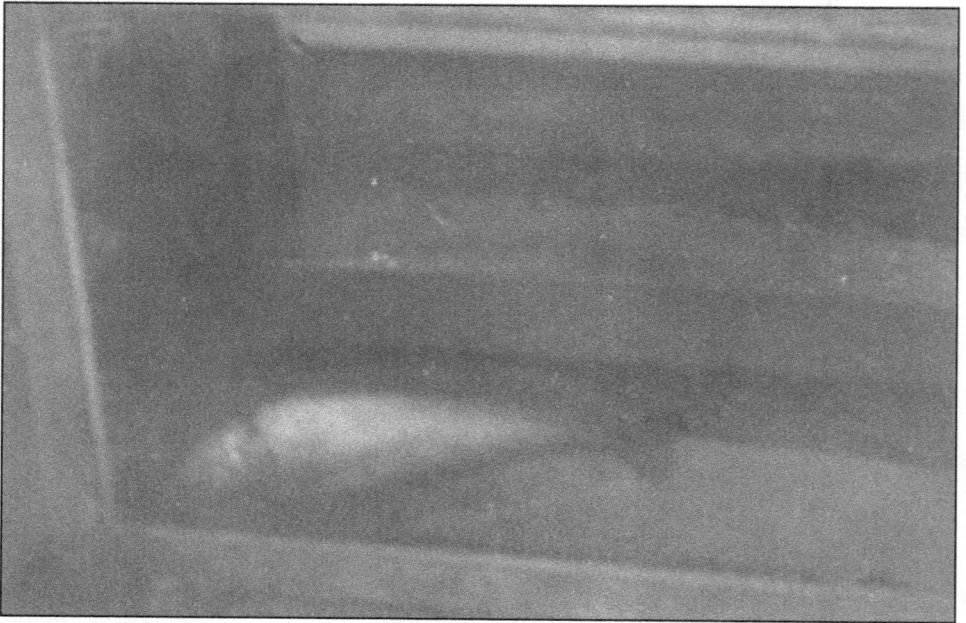

Figure 3.2g: *Notopterus notopterus*–Narcotized near positive electrode.

Figure 3.2h: Forced swimming to positive eletrode.

Figure 3.2i: Narcosis near positive eletrode.

Reaction of hybrid carp

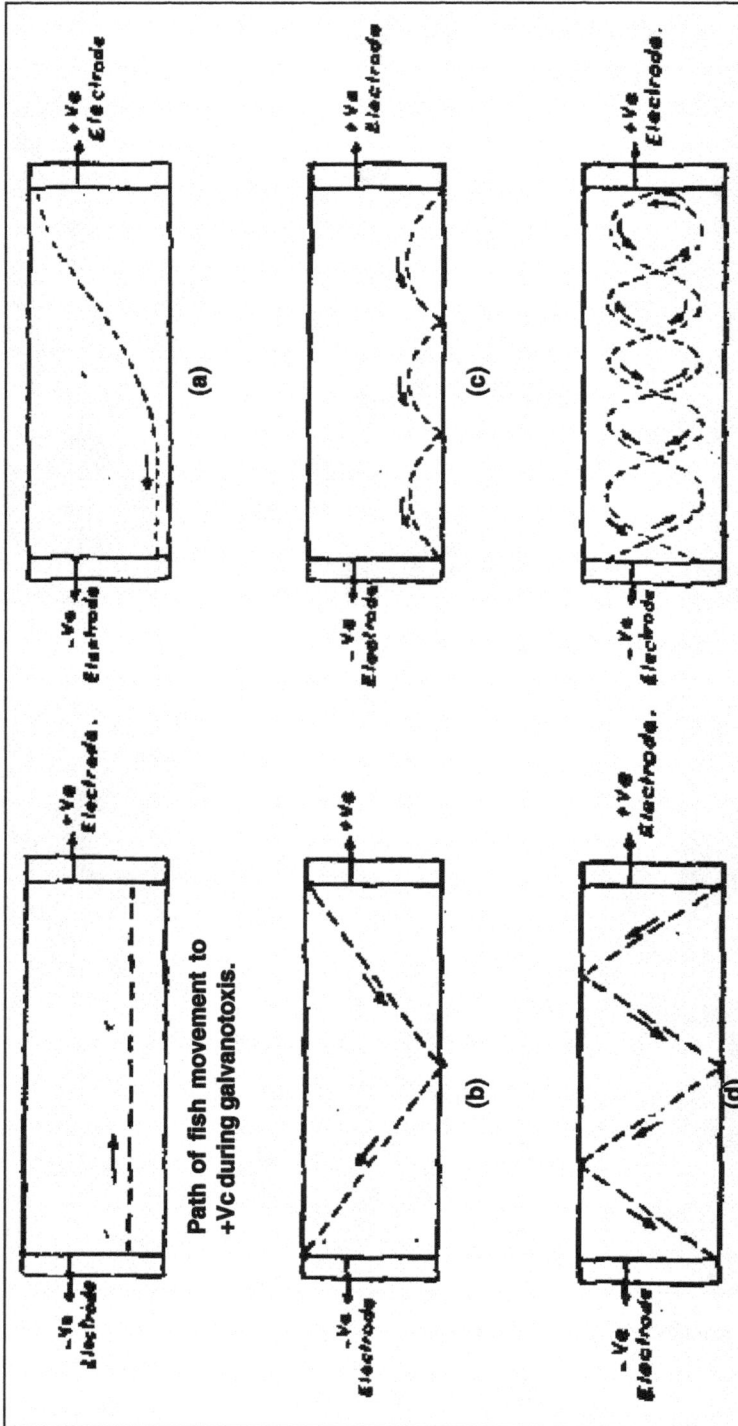

Figure 3.2]

(a-d) Path of fish movement to −Vc during galvanotoxis.

Path of fish movement to during galvanotoxis.

Chapter 4
Fishes in Electrical Field

Before understanding fish behavior produced by different electric currents, it is necessary to know their normal behavior in the observation tank or trough if not in their natural habitat. The different fish species behave differently under normal conditions both in their abode and under captivity. It is beyond the scope of study to observe their behavior in their natural habitat both in normal condition and in stress under external stimuli. Since most of the behavioral studies with respect to electrical shocks were studied in captivity, it is desirable to record their normal behavior before applying electrical stimulations.

In the tropical fresh water species, the normal behavior was studied with respect to small sized barb (*Puntius ticto*), a bear skin catfish (*Heteropneustis fossilis*) and a cichlid (*Tilapia mossumbica*) prior to subject them with electrical stimulation.

Normally *Puntius ticto* moved between the two extremities of the tank in a very coordinated manner. Their normal swimming speed was observed between 6 to 9.5 cm per second. Their dorsal fin dropped, the pectorals beat slowly and regularly and the cauldal fins showed lateral movements. Operculer beats were regular and were in the range of 238-253 per minute. The fish reacted very sharply to external stimuli. A little sound outside the experimental tank startled the fish, exhibited a fright reflex and the fish turned sharply away moving in the tank in a coordinated swimming movement. The fish even reacted in a similar manner to a shadow on the tank (Biswas, 1974).

Heteropneustis fossilis under normal resting conditions exhibited a very slow coordinated serpentine movement brought about by sequential alternate contractions of the myotomes on each side of the body for a short duration. The fish then settled in a comparatively darker area of the tank without any movement of their fins. A regular rhythm of gill movement of 162-174 beats per minute was noticed. It also reacted sharply to the external stimuli and quickly move away from the area of disturbance.

Tilapia mossambica under normal condition did not show much movement and rested on the bottom with slow undulation of pectorals in order to stabilize itself and prevent rolling on its sides. The rate of pectoral undulations were 10 to 40 per minute and their gill movement varied from 76 to 91 per minute. They were also subjected to fright reflex and reacted to external stimuli, but in a lesser degree than the other two species.

Fish Reactions in the Electric Field

Puntius ticto could feel the current and reacted depending on its initial position in the underwater electric field. The coordinated forward swimming of the fish retarded and came to a halt, when parallel to direction of current flow and facing the positive electrode. Occasionally a little backward movement was noticed as if scared to move towards the anode. The opercular beat slowd down to 34 to 42 per cent of the normal rate and the fish did not react sharply to other external stimuli.

The reaction was quickly followed by side ways jerks of the head, at times accompanied by short jerky movements mainly by the side ways movement of the tail and the head portion with an increase in field strength. At times the organism oriented itself almost perpendicular to field lines. The movement of fish still remained voluntary.

On facing the negative electrode, the fish was found to perceive the electric field by raising the median dorsal fin. The ventrals were spread out. Semi-circular movements accompanied with short jerks of the body (when parallel to field lines) succeeded by placing its body axis almost perpendicular to current lines without showing any further movement till the current was switched off. It neither reacted nor changed its position to external stimuli even when the fish body was touched with a glass rod.

Heteropneustis fossilis in the beginning perceived the surrounding electric field by stretching its barbells and vibrating its caudal fin and posterior portion of the ventral fin. The subsequent reactions with increased current strength facing positive and negative electrodes were similar to that of *Puntius ticto*, described in earlier paragraphs. Response to other external stimuli was reduced to considerable extent, and the rate of gill movement slowed down to 62 to 70 per cent of the normal.

Tilapia mossambica exhibited its response to the electric current by quicker undulation of pectoral fins facing the positive electrode. 42 to 55 per cent increase in pectoral beats has been observed under this stage of reaction. Facing the positive electrode, the pectoral, ventral and the dorsal fins were fully stretched and the organism oriented itself perpendicular to the current lines. Not only it became less responsive to other external stimuli, but also occupied the same perpendicular position even though the fish was placed parallel or in any position other than perpendicular to the current lines with the help of a glass rod.

The first experiment tried to identify the different reactions of fish produced by an electric field as a function of field strength and polarity. Several investigators tried to localize the nervous and muscle structures producing reactions, by sequentional neurosections, curarization and anaesthetics and determined the threshold for specific reactions. Others tended to apply electro-physiological theory, such as Pfluger's laws for the nerve excitation (Pfluger, 1859) and electrotonus laws modification of excitability (Charbonne Salle, 1859) to the fish situated in an electric field. The conclusions of these investigations were very different.

Theories Put Forth on Fish Behavior in Electric Field

The possibility of a galvanotropic reflex thought to be a reason for keeping the fish away from any direction other than anode (Nusenbaum and Falveea, 1961) is thought to be unlikely due to cinegrammatic evidence. The analogy of fish's reaction in an electric field in the same way as a Pluger's (1859) nerve-muscle preparation, the head and anterior part of the body as nerve and the tail as the muscle is physiologically impossible (Denzer, 1956). That fish would conduct current more easily when facing the cathode than anode and would prefer the least stimulating position (Blancheteau *et al*, 1961) and the existence of special nerve center in fish brain in the vicinity of vagus nerve (Kuroki, 19590 for galvanotropic reflex is not convincing as galvanotaxis occurs on a spinal fish. The encephalic reflex thought to be responsible for fish's preference for the anode (Balayev, 1981) is also not tenable as electrotaxis persists when encephalon is destroyed.,

Their theories remained pure hypothesis without success, since these theories assumed the simplification of the electrotactic mechanism. There is no reason to suppose that this is the case as the fish with complex network of nerves, nerve centers and muscles can be stimulated by the current independently or simultaneously. The anodic reaction theory is just the beginning and needs a study and a generalization of the phenomenon occurring in the whole fish during the action of a polarized current.

It appears to be only quasi-general agreement amongst most authors that electrotaxis is the result of anodic curvatures following each other on the two sides of the fish body (Le Men, 1980), although no explanation was available.

Physiological Basis of Fish Reactions in an Electric Field

The first physiological studies on electric fishing theory were made by Scheminzky (Scheminzky and Kollensperger, 1889), but the first statement of a coherent theory was not published until 1960s (Blancheteau *et al*, 1962; Lamarque, 1967). At this time mechanisms, such as, anodic galvanotaxis were not explained and studies on pulsed current were just beginning. This theory was verified by more recent methods, such as, nervous and muscular response records (Lamarque *et al*, 1975).

Several authors have tried to classify fish reactions in electrical fields. It is difficult to get general agreement on such classifications, as fish reactions change more or less according to fish species and current types.

A dynamic description of the reactions which occur can be found in the film, taxis, narcosis and electric tetanus in fish. (Lamarque, 1966) in which reactions are based on sequential neurosection. The fish species experimented were temperate fish species (Trout and eel).

Emperical studies on three tropical species, a barb, a catfish and a cichlid was made (Biswas, 1974) and the dynamic descriptions of their reactions in the electrical field are given below.

The reactions at current intensities beyond perception, encountered with forced involuntary movements of *Puntius ticto*. On facing the anode it rushed ahead towards positive electrode at a speed of 18 to 34 cm per second. The opercular beat increased to 58 to 71 per cent. Occasionally individuals jumped out of the water and a few manoeuvered to stay without much movement at right angles to the direction of current flow.

On facing the cathode fishes swam with short undulations in an elliptical path to a certain distance and maneuvered to turn to positive electrode moving fast towards it. The dorsal, pectoral and ventral fins were fully stretched out on facing the negative electrode and the gill movements were slowed down with irregular beats. The fish tend to stay perpendicular to current lines.

The fish across the field lines curved towards the anode, remained there till it receive a new stimulus of increasing intensity. The extremities of the body bent towards the electrode in the shape of an arch. Gill rate increased to 58 to 63 per cent of the normal. The fish did not react to any other external stimuli. The fish occupying a perpendicular position to current lines could not be moved from that position even though it was prodded from that position with a glass rod.

On facing the positive electrode, *Heteropneustis fossilis* moved fast towards it with long undulations of its body. This was effected by a slow contractual waves of high amplitude passing down the body from head to tail. The speed of this forced movement was 42 to 54 cm per second. The opercular movement slowed down to 17 to 58 per cent of the initial rate. A few individuals oriented themselves perpendicular to lines of current conduction and did not move from that position.

Rapid contractual waves of low magnitude passed down the body from head to tail resulting in the jerky swimming of fish with its head turned a little towards the anode, when the fish head pointing towards negative electrode at the start. On maneuvering to turn its head towards the positive electrode, it reached the electrode with long undulations of the body. This forced movement continued till the fish could orient itself either perpendicular to field lines or became perpendicular to the bottom plane.

Head and tail of fish body curved towards the positive electrode, it being perpendicular to the field lines. The extent of curvature increased with the increase of potential. The organism also did not react to any other external stimuli at this condition like *Puntius ticto*.

A higher rate of pectoral undulations (up to 320 to 450 per cent per minute) from normal rate, followed by tremor of the body and expansion of fins were observed in case of *Tilapia mossambica* when its head pointing towards the anode. It also invariably attempted to orient its body axis perpendicular to the lines of current conduction. On facing the cathode it exhibited jerky movements with a curvature of the body resulting in short circular movements in the field. The rate of gill movement slowed down to 15 to 40 per cent. Being across the field lines, an increased pectoral beats up to 155 to 230 per cent accompanied by bending of body extremities towards the positive electrode was observed. The extent of curvature of the body depended on the field intensity. No response to any other external stimuli under this condition was noticed.

On increasing the current intensity further, the fishes of three varieties lost their stability and either rolled over on one side or on its back incapable of any voluntary movement. The body of the fish remained motionless with expanded fins, either near the anode or in the field turning its head towards the anode. The gill movement were irregular and the amplitude of each opercular beat became larger initially, and later there were sudden rapid beats of the opercle with an interval of few seconds. Fishes of all the three species exhibited bending of their body extremities towards the anode when the fish was lying perpendicular to field lines. The curvature of the body was more and sustained after the rise of field intensity and no movement on the part of organism was noticed.

Body pigment of *H. fossilis* darkened to deep yellow from a ligher shade at this stage. The chromatophores of *T. mossambica* deepened to black with reddish tinge in the fin extremities from grayish white, while the fish was heading on to the anode. But on its turning the head towards the cathode, the grayish white color of the body faded to white during the current flow.

These observations were made in large tanks in which a homogeneous field (constant current intensity) was created between two electrodes. The response of fish was identified in threshold current density of the field in which the fishes were subjected. These parameters were related to the voltage gradient through the conductivity of the water;

Voltage gradient = Current density/Water conductivity

Current density – Voltage gradient squared X Water conductivity

The first reaction observed, irrespective of the orientation of temperate fish in the field, was a quivering motion of the body or dorsal fin, which occurs in the perception zone around the electrode. This preliminary reaction does not occur if the DC field is pure or applied slowly. If the circuit is closed abruptly, the threshold for these reactions was 20-80 mV/cm depending on the fish species.

In tropical barbs (*Puntius ticto*) of 61 to 100 mm total length. Catfishes (*Heteropneustis fossilis*) of 106 to 170 mm and cichlids (*Tilapia mossambica*) of 91 t0 200 mm total length, the mean threshold current densities required to initiate the first reaction were 0.04 to 0.08 delta for *P. ticto*; 0.06 to 0.15 delta for *H. fossilis* and 0.15 to 0.45 delta for *T. mossambica* in slowly rising DC field (Biswas, 1974). The current density was calculated as the field strength in micro-ampere passing through an unit area of one square mm in a homogeneous field.

The temperate fish after the first reaction moves into the effective zone of the electrode (may not be a homogeneous field). The responses observed correspond to the increase voltage, current density or proximity to the anode and depend on the direction the fish is facing with respect to the anode.

Fish Facing the Anode

Inhibited swimming occurs at approximately 120 mV/cm. The normal swimming of the fish is retarded as can be clearly seen from the absence of swimming vortices.

As the voltage gradient increases towards 150 mV/cm, the fish rushes ahead, swimming strongly with very large undulations of the body. This first swimming towards the anode (forced swimming) is not an anodic galvanotaxis, but a component of this reaction, which only occurs when the fish is facing the anode. Galvanonarcosis appears at 340 mv/cm (the current producing tetanus when the fish is facing the cathode). The body of the fish becomes motionless and its muscle relaxed.

At 800 mV/cm, the fish which was in a state of galvanonarcosis begins to swim again, but in an unbalanced manner. This is the second swimming to the anode (pseudo-forced swimming) which is often accompanied by tetanus (eel). When the current value producing this reaction is reached the fish can not escape the anode direction. Anodic galvanotaxis is now complete as the fish is now obliged to go towards the anode.

Anodic tetanus of muscular origin occurs at values above 1000 mV/cm (Lamarque, 1967). As the reaction occurs after curarization, a drug application which prevents the transmission of the nerve impulse to the muscle. This tetanus is the result of direct excitation of the muscle by the current.

Fish Facing the Cathode

With fish facing the cathode the situation is reversed in term of polarity and nerve excitation. Increasing voltage induces two specific reactions resulting in eventual orientation of the fish towards the anode.

Facilitated swimming- cathodic galvanotaxis at approximately 116 mV/cm (the value producing inhibited swimming when the fish is facing the anode) the fish tends to swim away from the anode with short fast undulations of the body. If the voltage gradient is increased to 150 mV/cm, the fish turns itself to face the anode. This is the mechanism of half turn towards the anode.

If the fish does not undergo the half turn towards the anode, and remains facing the cathode, a rapid increase in voltage gradient to 350 mV/cm will result in cathodic tetanus of nervous origin (at the same threshold value, which produces galvanotaxis in fish facing anode), when the body is tetanized and quivers intensely, the effect disappears after curarization indicating its nervous origin, At higher voltages (more than 1000 mV/cm) this is followed by cathodic tetanus of muscular origin, evident by the absence of quivering.

Fish Lying Across the Field

In this orientation only one response, anodic curvature is observed. This occurs at potential above 350 mV/cm. As the reaction disappears after curarization, the response is due to nervous excitation and not direct muscle stimulation.

Physiological Mechanism Involved in Fish's Reaction in a DC Field

The Neuro-Muscular System

The nerve element is composed of a body cell, nerve fiber and terminal arboration (Figure 4.1).

Elements are positioned, in series and in parallel and form chains through interconnecting synapses, where the transmission of the impulse is elaborated. A sum of elements makes a nerve, which can be sensory or motor. Sensory nerves drive the

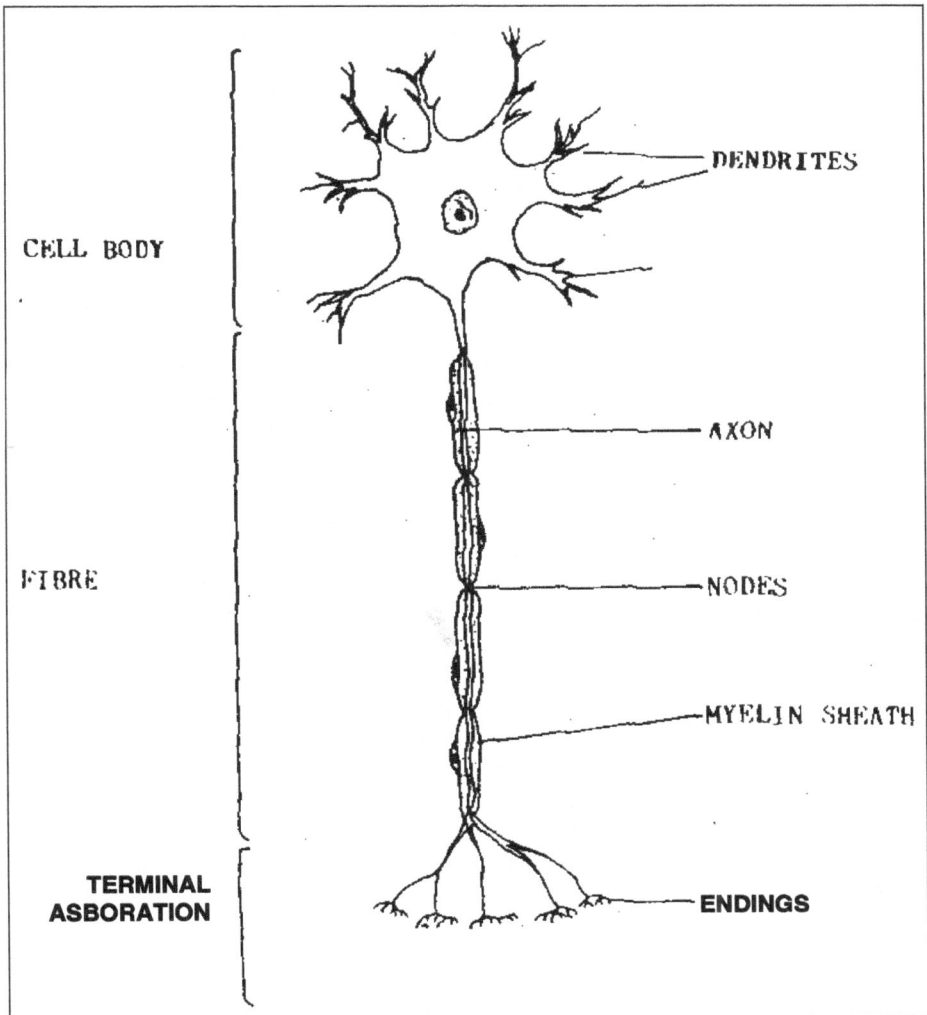

Figure 4.1: Diagrammatic representation of a nerve element.

peripheral excitations to the brain or medulla where the reflexes are initiated. These reflexes are transmitted to the muscles via motor nervous system (Figure 4.2).

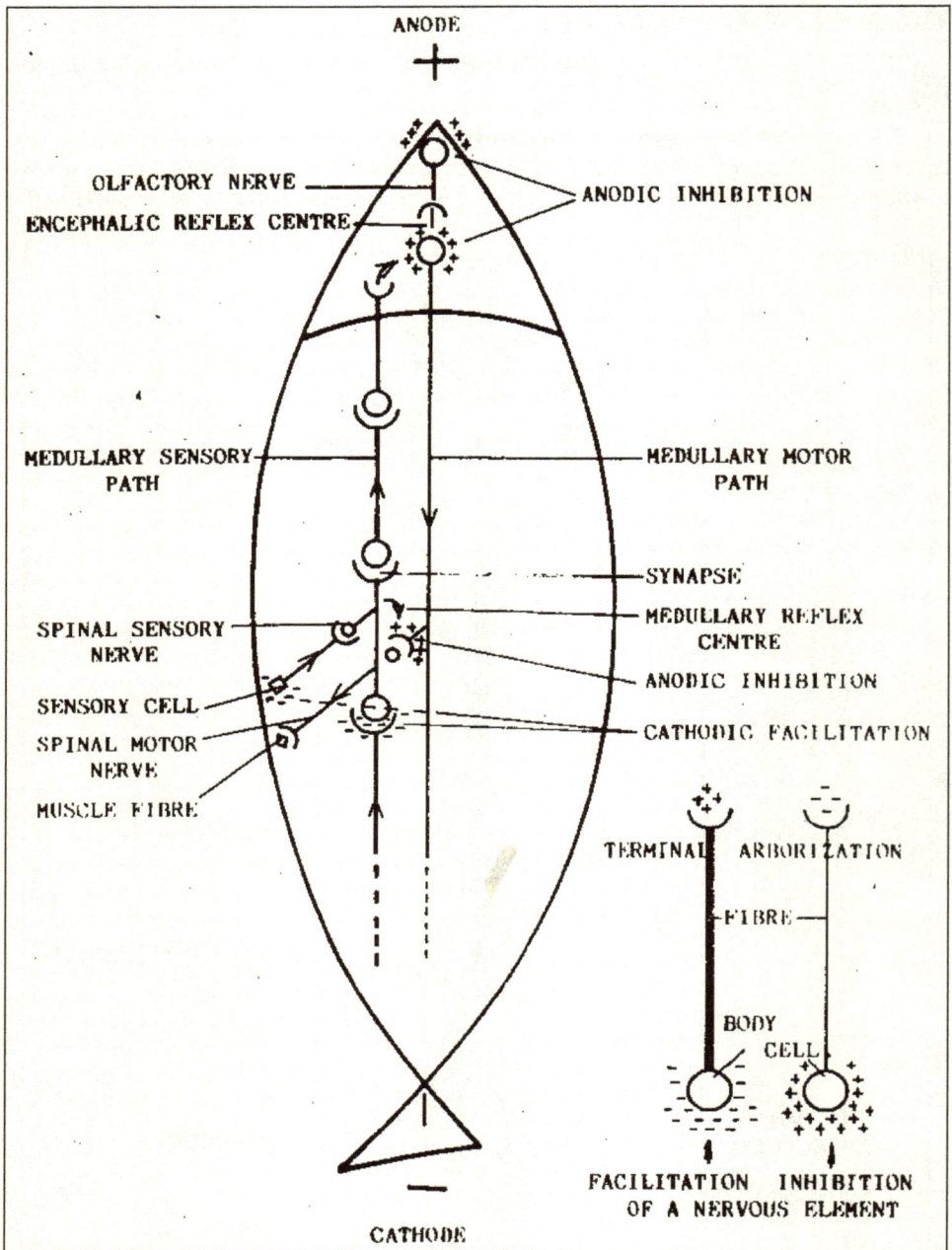

Figure 4.2: Diagram of the nervous pathways in fish and representation of the mechanism of facilitation and inhibition.

In fish, the sensory and motor spinal nerves are oriented in the body at about 45 degree to the longitudinal body axis, and are symmetrical on the two body sides (Figure 4.2). Muscles are composed of fibers which tend to be very short in fish (Nursall, 1956).

Electrotonus : DC Action on the Body Cells

Pure DC acts on the body cells (electrotonus) and muscle fibers, but not on the nerve fibers. Pulsed DC acts on the nerve fibers and muscles (Pflugger's Laws). Under the action of DC, the laws of electrotonus apply (Charbonne Salle, 1881). According to these laws, the body cell is either facilitated or inhibited by DC. Facilitation means an increase in the excitability of the cell and inhibition a decreased excitability. If a body cell is facilitated, it will transmit its stimulation to the fiber either producing an increase in the reflex impulses coming from a higher pathway or for higher voltage values, produce a direct stimulation of the fiber (Figure 4.2). If inihibition occurs it will diminish stimulations coming from a higher pathway or suppress these stimulations producing an anodic inihibition or block (Figure 4.2).

Mechanism behind the Reactions

A body cell is stimulated by DC when the cathode is situated on the body cell side with regard to its prolongation. The body cell is inhibited when the anode is situated on the body cell side. A variation of voltage is not necessary to produce these two effects. Nevertheless, it is possible to produce them with pulsed current through summation (temporal) of individual pulses which act on the body cell as DC. DC stimulation threshold of a body cell producing a response of the nerve is higher than a direct stimulation threshold of a nerve fiber by pulsed current. Finally DC directly stimulates muscle fibers which respond in a sequential manner.,independently of the current polarity, for higher voltage values than pulsed ones.

Rushton's Law

The influence of the length of the nerve and its position in the electric field on the threshold for stimulation was first pointed by Rushton (1927) and elucidated by Lamarque and Charlon (1973).

The threshold for stimulation of a nerve or muscle situated in an electric field decreases as a cosine function of the angle each forms with the current lines (Figure 4.3).

Thus, when a fish is parallel to the current lines, a very small changes in angle will produce a large variation in threshold. The position of the fish in the electric field therefore, has a considerable effect on the threshold values for stimulation (or inhibition) of the excitable structure. The first to respond will be those parallel to the current lines, the last those perpendicular. When a fish is swimming in an electric field, the excitability of structures is being constantly modified by the changing orientation of its spinal nerves and medullary nerves in the field.

The threshold stimulation (or inhibition) of a nerve or muscle situated in an electric field decreases as a function of its length (Figure 4.4).

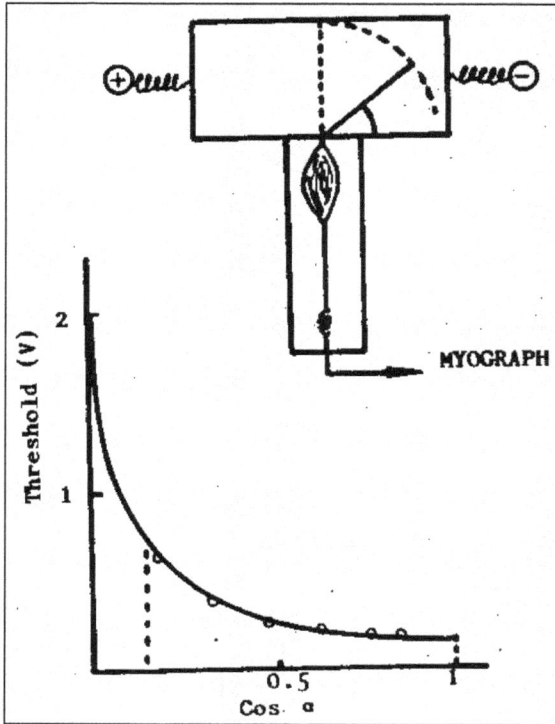

Figure 4.3: Rushton's first law, Influence of nerve orientation in an electric field on the threshold of excitation.

Short structures are therefore, less easily stimulated than long ones and from a practical point of view, the threshold remains stable for nerves larger than 4 cm (Figure 4.4).

Mechanism of Anodic Galvanotaxis

Galvanotaxis is an obligatory swimming artificially induced by DC, independently of the direction taken by the fish. The galvanotaxis can be anodic or cathodic. Anodic galvanotaxis is swimming produced artificially by the current compelling the fish to approach the anode. This reaction is relevant in the case of a non-homogeneous field that exists in the field situation, where a fish swims in an electric field of increasing value. The best result of anodic galvanotaxis (electrotaxis from pulsed currents) in a non-homogeneous field is shown by moving the anode around in the water, the fish will follow the electrode in the direction of its motion.

When the initial starting orientation is facing the cathode, the sensory nerve body cells in front of the encephalon are stimulated (Figure 4.5A) producing a swimming reflex. During the first swimming undulations towards the left (chosen arbitrarily) in Figure 4.5, the body curves itself towards the position shown in

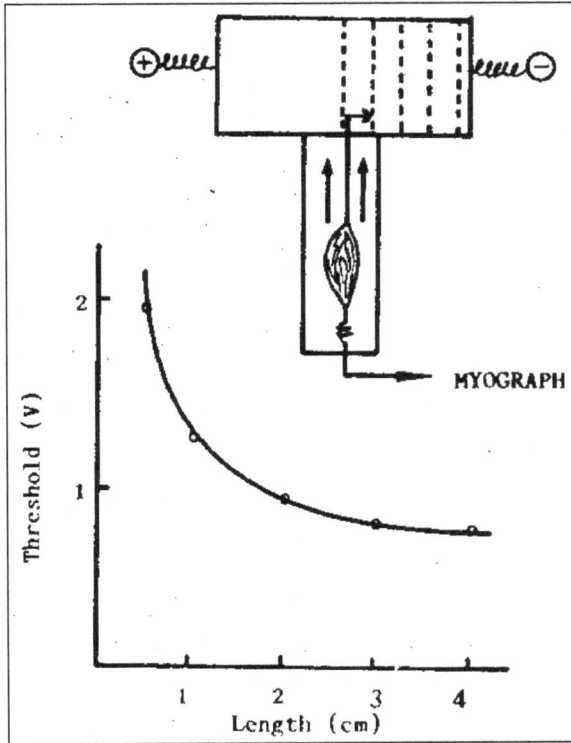

Figure 4.4: Rushton's second law- Influence of the length of the nerve situated in an electric field on the threshold of excitation (from Lamarque and Charlon, 1973)

Figure 4.5B. Spinal motor nerve stimulation becomes asymmetrical with those on the left side of the body being facilitated to a greater degree. This higher facilitation on the one side of the body leads the fish to persue its curvature towards a transverse position in the field (Figure 4.5C)- anodic curvature. In this position, the spinal nerves on the left side of the body (near side to the anode) are facilitated and those on the right side are inhibited. One sided stimulating of the body tends to induce the fish to pursue its half turn. In this position (Figure 4.5D) spinal nerves on the right side of the body are strongly inhibited, but no inhibition occurs on the left. Thus reflexes on this side of the body are conducted better and the fish completes its half turn towards the anode (Figure 4.5E).

The fish is now facing the anode. If the voltage value is low, first swimming towards the anode can be induced making the fish approach the anode and be submitted to a higher current value.In the stage of first swimming the fish can escape the anode direction, but if it reaches the current density zone where the second swimming to the anode is produced, it is no longer able to escape the anode direction. This is because the asymmetrical excitation of the spinal nerves brings it back to the anode each time it stays away from this direction. In this second swimming, the reaction is the result of spinal motor nerve participation as curarization suppress the

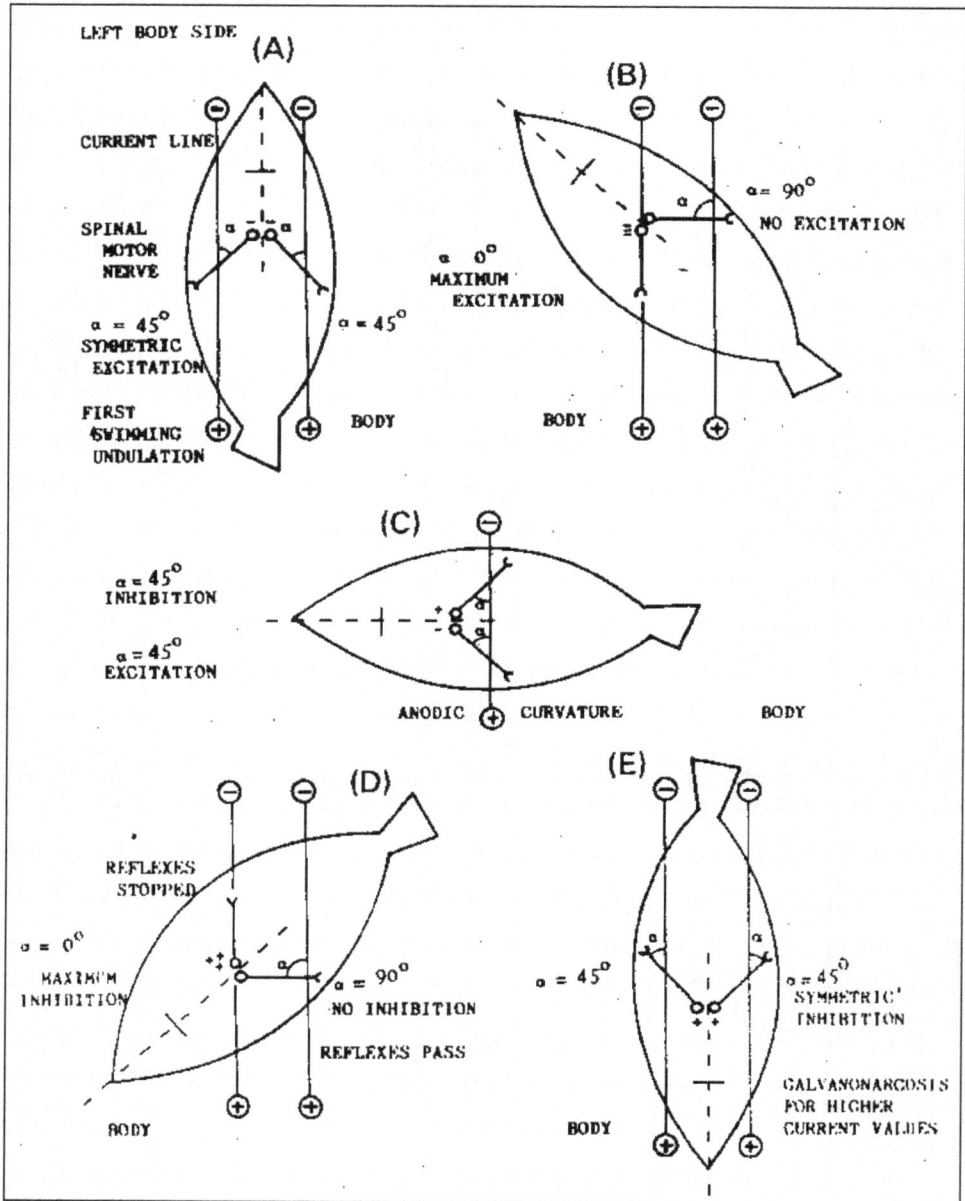

Figure 4.5: Mechanism of half-turn towards the anode observed when fish in a dc field are initially facing the cathode.

action, but spinalization (destroying the spinal cord) does not. Anodic galvanotaxis is thus achieved.

When the fish was initially facing the anode, as soon as the fish perceives an electric field with a higher voltage value than that required to induce inhibited swimming, the fish swims. This is the mesencephalic brain reflex, forced swimming (Blancheteau *et al*, 1961) called in the present terminology first swimming towards the anode. The fish thus approaches the anode and enters the second swimming zone, as considered above.

The undulations of the first swimming to the anode is very ample, because they benefit from the facilitation-inhibition mechanism on the spinal motor nerves at each change of direction produced by the undulations. This adds its effect to the normal swimming movements. In fact, whatever its position in the field, the fish is constantly attracted towards the anode through asymmetrical excitation-inihibition of the spinal nerves. It is only a tendency at low current values, but at higher values the fish is compelled to approach the anode. To sum up, anodic galvanotaxis is made up of three reactions; half turn towards the anode, first swimming and second swimming towards the anode.

Mechanism of Muscular Tetanus

When the fish is closed to the anode, the fish can be tetailized in DC field. At this time the most muscle fibers reach their excitation threshold. As soon as the fish moves away from electrode, it recovers and returns to the second swimming zone for the whole mechanism to be repeated.

Responses in Pulsed Current Field

The reaction of fish to pulsed DC is quite different from DC. The mechanism of pulsed DC is extremely complicated and made more so by the dynamic behavior and unlimited types of pulsed DC available. In the case of square-wave pulses at a frequency of 100 Hz and 1 ms pulse duration, the following reactions produced by DC do not occur, with this type of pulse;

- ✰ Inhibited swimming,
- ✰ Galvanonarcosis,
- ✰ Second swimming towards the anode.

When the fish is facing anode, normal fish swimming with large undulations is replaced by swimming with short, fast undulations. In comparison to DC, the threshold is considerably reduced. For example, in eel from 310 mV/cm with DC to 81 mV/cm for pulsed current (Lamarque, 1976a). Tetanus occurred at considerably lower threshold (= 165 mV/cm) than for DC (1400 mV/cm) without any other intermediate reaction.

If the fish is oriented across the field, anodic curvature occurs at 136 mV/cm instead of 350 mV/cm in DC. When the fish is facing cathode, a first swimming occurs at the same threshold as that when facing the anode, but it is followed by a half-turn towards the anode. However, this half-turn is not as efficient as with DC. Thus with pulsed current there is not a complete half turn nor a true anodic electrotaxis.

Physiology of Reactions in Pulsed Current Field

According to Pfluger's principles to be directly stimulated by a pulse a fiber heads;

☆ A minimal voltage (threshold),

☆ A minimal current duration (useful time) and

☆ A variation of voltage of voltage (steady DC does not stimulate the fibers)

The fiber will respond;

☆ At closing the circuit (closing the circuit reactions)

☆ At opening the circuit (opening the circuit reaction) but only for a current value higher than the threshold value for the closing circuit reaction

☆ If there is a negative correlation between the threshold and the current establishment slope,

☆ If there is no current polarity effect on the fiber.

It can be assumed from these laws that from the threshold, a pulse of sufficient duration (superior to the useful time) could excite simultaneously all the excitable structures of an organism situated in an electric field. As a result, sensory and motor nerves and muscle produce simultaneously direct motor end reflex responses and tetanus independently of the current polarity.

In the majority of work on the electro-physiological effects of pulsed DC, attention has focused mainly on current frequency and duty cycle. The most important parameters of a pulse is its duration which falls into three categories.

1. That necessary to produce the closing circuit reaction on the fiber, that is, the useful time.

2. A protracted current which induces a DC response on the body cell; and

3. A duration and voltage sufficient to produce, the first closing circuit reaction on the fiber followed by an opening circuit reaction.

Thus, if the difference in stimulating effect between closing and opening of the circuit is sufficiently great, as occurs with a current which increases slowly and falls abruptly, when the circuit is opened fish are only weakly stimulated around the anode and swim towards the cathode. Under these circumstances, the slow rise has no stimulating effect because the nerves become accustomed to the current (Lamarque, 1967 a b) and the abrupt fall produces a stimulation of sufficient force to induce the fish to swim towards the cathode. Conversely, if the current rise sharply and falls gradually, it does not produce movement towards the cathode, but an unmistakable anodic reaction. This clearly demonstrates that pulsed direct current should have a suitable form if an anodic response on fish is desired.

For a precise analysis of pulsed current it would also be necessary to take into account such parameters as synaptic delays and refractory periods. For example, in a chain of pulses, all the pulses falling on a fiber during refractory period will have no effect. Thus the pulsed current mechanism is complex and even more so as fish behavior dynamic and the number of current types are unlimited.

Mechanism of Anodic Electrotaxis

Again there are differences in response depending on the initial orientation of the fish. When the fish is facing the cathode, at low current values (80 mV/cm) all the sensory and motor fibers are more or less stimulated and the fish swim. As with DC, fish body curves towards one direction and reaches the anodic curvature position. In this position spinal nerves on one side of the body are more excited than those on the other. From this position pulsed current mechanism is quite different from that observed for DC. In fact, when the fish is across the field, spinal nerve stimulation of the two body sides is symmetrical, as there is theoretically no polarity effect with a pulse. In this position anodic curvature fully occurs showing a difference of body excitability, between the two sides. The reasons behind this are not fully understood. As the fish turns further towards the anode the stimulation again becomes assymetrical but differently from DC as the right side of the body is stimulated to a greater degree. Under these circumstances the fish does not show a trend towards the anode but again towards the transverse position in the field. With pulsed current there is no complete half turn nor a true anodic electrotaxis. If it reaches the anode, it is because the inertia produced by the fish in the anodic curvature carries it around.

When a fish is facing the anode, at the same current value that stimulates swimming when facing the cathode (80 mV/cm), the fish swims to the anode (first swimming towards the anode). However, if it is oblique in the field it will not show a tendency to swim again towards the anode, as with DC. The trend will be to lie across the field.

Mechanism of Tetanus

At a very low voltage (160 mV/cm) in comparison to DC, the fish is tetanized. All the excitable structures of the fish are simultaneously stimulated independently of polarity. Under normal electric fishing conditions the fish has the possibility to turn more or less toward the anode, to swim a little and then be tetanized often before reaching the anode.

Alternating Current

With this current type it is important to note that the current direction is changing every half cycle and there is no polarity effect on the organism. Thus the fish faces cathode and anode successively as many times as the current alternates.

In a homogeneous low-density field, if the fish is facing one electrode, it will swim towards it as there is a general stimulation of the body similar to pulsed current. This reaction is termed oscillotaxis. The mechanism occurs because, at all frequencies, the reaction threshold is lower when the fish faces the cathode, and in an alternating field the fish will always react as if it is in such a position. Under these circumstances, there is a general stimulation of the motor nerves by the negative changes. The reflex excitation, which usually occurs when facing the anode does not take place during the positive half of the cycle as the reaction threshold is high. The optimal frequencies for inducing oscillotaxis in European eel and brown trout are 200 and 100 Hz respectively (Lamarque, 1967 a).

Figure 4.6: Electrotactic effect in a fish farm pond by (a) full-wave rectified tri-phase AC at 300 V and (b) square wave at 100 Hz at 200 V. Water conductivity – 300 micro S/cm.

After several runs in each direction the fish adopt a transverse position in the field, transverse oscillotaxis, where it remains. In this condition stimulation on one side of the body is always superior which, according to Rushton's Law, makes the fish adopt transverse position. In this position there is no anodic curvature because stimulation remains symmetrical on the two sides of the body, there is no polarity effect with AC. If the fish tries to escape, it is forced to return and eventually it succumbs to tetanus. For a small increase of voltage the fish is tetanized on the spot.

Under field conditions, a fish which faces the electrode at a low current value swims in this direction, reaches more or less by inertia and is tetanized. For fish not oriented in this direction, it will swim away from the electrode if the current density is low or will be tetanized if it is high.

When Tilapia was exposed to AC, their reaction is not typical (Lamarque, 1975 a). Eighty per cent rise to the surface and remain there swimming weakly. This reaction is potentially useful for catching these fish and any other species that exhibit this reaction. The cause of this response is unknown.

Comparison between DC, Pulsed Current and AC Reactions

The range of habitats to be fished varies considerably and the responses of fish to different current types are important in the choice of equipment in the field situation. DC has good anodic galvanotaxis and induces tetanus only in the near vicinity of the electrode.

Pulsed DC has poorer anodic electrotaxis and tetanizes further from the anode preventing some fish reaching the electrode.

AC has no electrotaxis and fishes are tetanized at greater distances from the electrode than pulsed DC or DC.

The different electrotactic and tetanizing effects of DC and pulsed DC (100 Hz, I ms duration); (Figure 4.6). With DC, fish are gathered, immobilized close to the anode, but are more dispersed for pulsed DC. Moreover, with pulsed DC, some trout in the experiment had broken vertebrae, whilst those with DC had no injury.

The relationship between first swimming threshold and tetanus threshold expresses this variability in anodic electrotaxis. In general, the most useful current is one which attracts from a great distance and stuns close to the anode, that gives the highest ratio value (Lamarque, 1976 a).

With AC the ratio would have been considerably lower, but has not been measured.

Ratio between Threshold Potential to Evoke First Swimming and Tetanus

Current Type	Ratio
DC	4.54
Pulsed DC 100 Hz	2.03
Pulsed DC 400 Hz	2.49

Chapter 5

Empirical Studies on Cold Water Fishes in DC and Pulsed DC Field

Reactions and Thresholds

Cold water fish species, like, *Salmo irideus, Idus melanotus* and *Cyprinus carpio* were studied for their electro-physiological behavior in cold water (15-16 degree Celsius) in DC and pulsed DC. Pantostat 523, an electronic impulse transmitter of 60 W capacity, produced continuous DC and impulsed DC of variable frequencies. The homogeneous field created in the experimental tank of 200 cm X 54 cm X 25 cm was used to study the behavior of fish in the electric field. The electrical conductivity of the water was measured by Siemen's conductivity meter and the wave form and frequency of impulse DC was checked through a cathode ray oscilloscope.

A fresh fish was used separately for each observation and the conductivity of water in the tank was raised by dissolving sodium chloride.

The intensity of electrical field in the tank was raised from zero till the fish exhibited increased gill movement, expansion of dorsal, pectoral and caudal fins accompanied by tremor of the body and occasional jerks (first reaction). With the rise of current intensity beyond these visible reactions, a directional movement of the fish to the electrode (galvanotaxis) was observed when the fisk lay parallel to the direction of current. But when the fish did not lie parallel to the direction of current, it first turned its head towards and then move to the electrode. On further rise of current intensity, the fish could no longer move out and lay on its side when the intensity reached its threshold value (galvanonarcosis) (Biswas, 1971).

Since the electrodes used in the study were plane and parallel at a distance of 200 cm, an uniform electric field was produced in the tank water. The density of electric field (delta) was calculated as the current in micro-A passing through a unit area of cross section (square mm) Current density was found to be the most critical factor in producing forced directional movement of fish. With the increase in current density the fish rapidly swam towards the anode when the head was facing the anode and took a straight path.

The fish took an elliptical path and moved violently towards the cathode during galvanotaxis when the fish head was pointed towards the negative electrode. The fish in transverse position to the current direction exhibited irregular movements between the electrodes during galvanotaxis. With further rise of current density the fish sank to the bottom of the tank and lay on its side ceasing all voluntary movements.

Houston (1949) observed the tendency of fish to place themselves with heads pointing towards the increasing potential. Cattley (1955) described the reactions of fish in three definite stages which corresponds with first reaction, galvanotaxis and galvanonarcosis described earlier.

Threshold current densities for different reactions varied inversely with the length. Not only the absolute current density decreased with increase in fish length, but also the ratio of current densities for a particular reaction decreased with the increase of fish size. *Phoximus laevis* of median length 19 mm exhibited a mean ratio of current density of 1:17:25 for the tree reactions, while corresponding ratio for fish of 66 mm length was 1:10:17 (Schemanzky, 1938).

Both *C. carpio* and *S. irideus* exhibited the three reactions with increase in current densities. The ratio of threshld current densities for these reactions were found to be 1:3:10 and 1:1:4 in *C. carpio* lengths 91 mm and 230 mm respectively. In the case of *S. irideus* of lengths 140 mm and 250 mm, the corresponding ratios were 1:3:10 and 1:2:5 respectively. (Biswas, 1971).

Water resistivity has a very great effect on the current density required for producing electrotaxis and paralysis (Harris, 1953).

It has been observed that threshold current densities required to exhibit a specific reaction by *Idus melanotus* and *Cyprinus carpio* increased with the increase in conductivity of water, irrespective of size of fish. Though the optimum current density for a particular reaction varied directly with the increase in conductivity of water, their ratios in waters of different conductivities remained constant *viz*; 1;2:5. (Biswas,1971).

Pulsed Current

Houston (1949), using impulse DC of triangular wave shape having frequencies of 2 to 20 pulses per second with impulse duration of 2 ms could regulate the size and type of fishes caught by varying the pulse rates. Best effect was obtained by Kreutzer (1951) with impulse current of sudden increase and slow decrease and duration of 2 ms, when small fishes showed best results with 20 shocks per second and larger fishes with 2 shocks per second.

The optimum current densities required for the three reactions of *Salmo irideus* were found to vary inversely with the rise of impulse frequency, when pulsed DC of

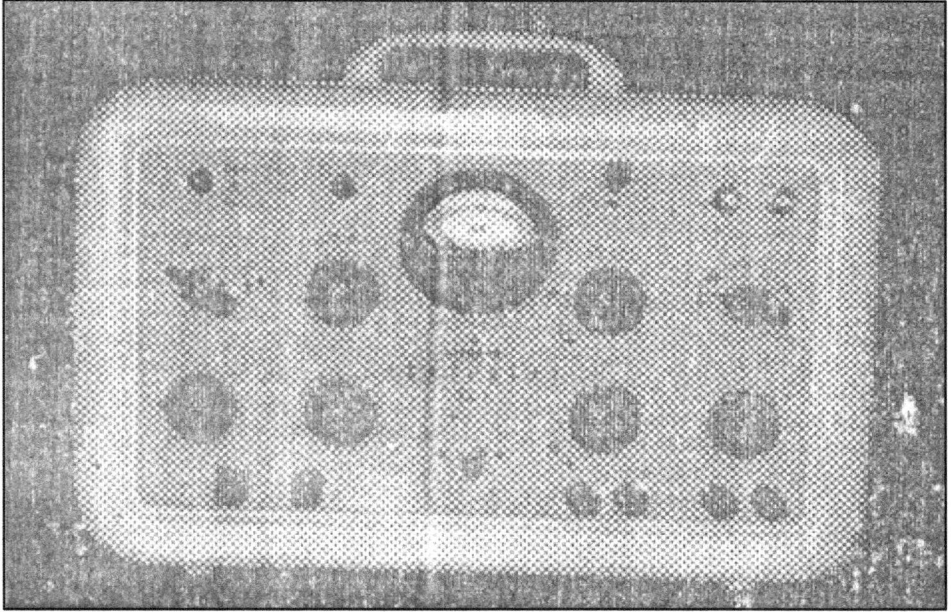

Figure 5.1a: Siemen's Pantostat 523.

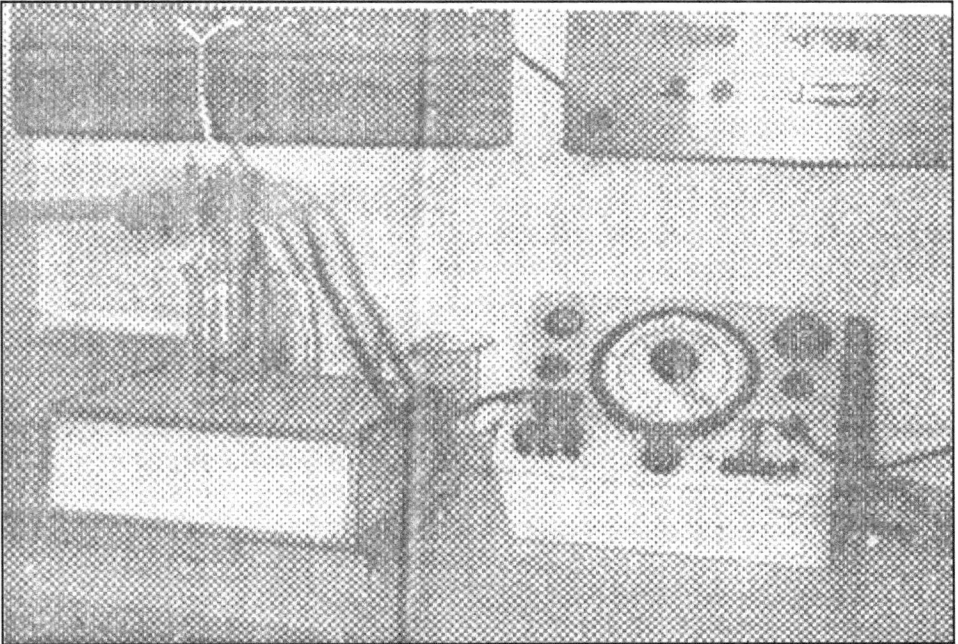

Figure 5.1b: Siemen's conductivity meter.

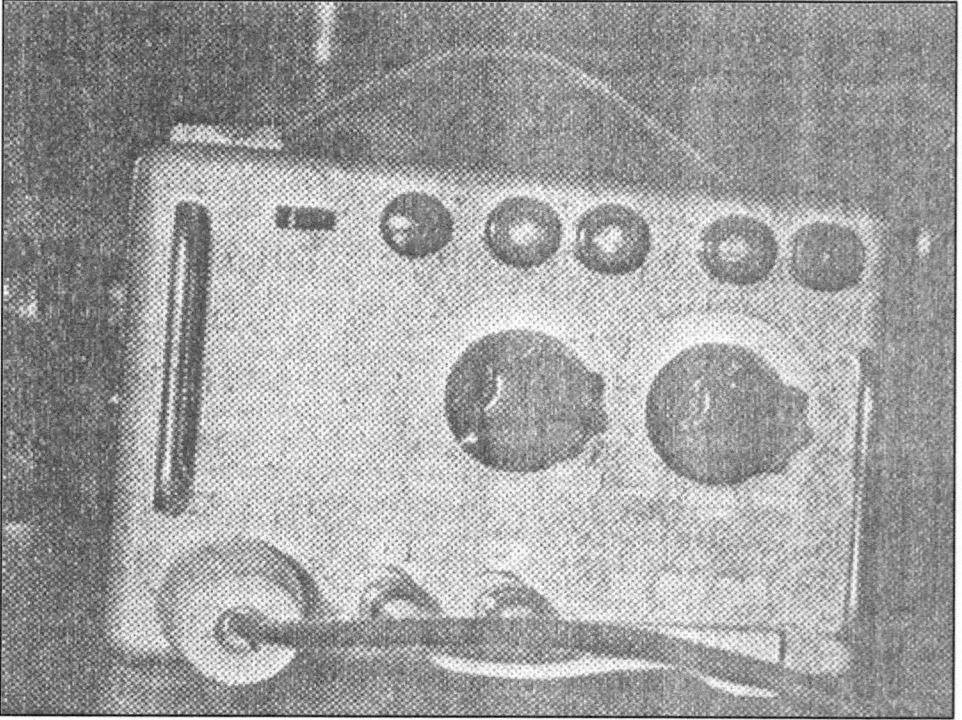

Figure 5.1c: Showing electronic impulse transmitter.

square wave form having frequencies between 26 and 48 per second with varying impulse duration from 1 to 18 ms and constant pause duration of 20 ms was used (Biswas, 1971).

While observing the behavior of *Carassius auratus* under the influence of low frequency electric shocks, it was noticed that the mean value of electric current intensity increased with the increase of frequency of impulses.

The threshold current densities required for first reaction in *Idus melanotus* of different lengths increased with decrease of impulse frequencies in water having Siemen's conductivity of 2×10^{-4}/ml. In higher conductivity of 7.6×10^{-4}/ml, the values rose with decrease of impulse frequencies in fishes of similar size grades. The threshold values of current densities for the second and third reactions varied inversely with the rise of impulse frequency irrespective of size of fishes and conductivity of water (Biswas, 1971). Morgan (1953) observed that the peak current value reached during a pulse was an important factor influencing the response of fish to the current.

Hosl (1955) observed that impulse current of 1 ms duration and interruption of 9 ms has greater physiological effect on fishes than DC or AC of 50 c/s.

Salmo irideus of 131 to 200 mm length showed 100 per cent anodic movement during galvanotaxis and galvanonarcosis in water temperature of 17 degree Celsius in impulse frequency of 82/second. 100 per cent of anodic galvanotaxis was observed at 11 degree and 15 degree Celsius. 50 per cent cathodic galvanotaxis occurred at 6 degree Celsius. Oscillotaxis and un-specific movements of *S.irideus* were found at a water temperature of 3 degree Celsius which did not exceed 14 per cent (Biswas, 1971).

Using impulse DC of square wave form *S. irideus* of 141-200 mm, exhibited 100 per cent anodic galvanotaxis and galvanonarcosis in impulse frequency of 82/second with impulse duration of 5 ms and pause duration of 7.2 ms in water conductivity of 2×10^{-4}/ml. Similar reactions of fish were noticed in higher water conductivity (10×10^{-4}/ml) and impulse frequency of 82/second when 100 per cent anodic galvanotaxix and galvanonarcosis occurred.

The pulse threshold for *S. irideus* of 141-200 mm in impulse DC of rectangular wave shape was 34/second with impulse duration of 9 ms and pause duration of 20 ms irrespective of conductivity of water; while in continuous DC fishes exhibited 82 per cent cathodic galvanotaxis and 100 per cent anodic galvanonarcosis in waters having conductivity of 2×10^{-4}/ml.

S. irideus in impulse DC of square wave form showed reduced anodic reaction both during galvanotaxis and galvanonarcosis in impulse frequencies both higher and lower than 82/second irrespective of conductivity of water. Using pulse DC of rectangular wave shape, the optimum impulse frequency for anodic reactions was found to be 34/second in *S. irideus* of 111 to 210 mm irrespective of conductivity of waters. Reduced anodic effect was observed in impulse frequencies both higher and lower than 34/second (Biswas, 1971).

Idus melanotus of 171 to 200 mm exhibited 100 per cent anodic galvanotaxis and 80 per cent anodic galvanonarcosis in water having conductivity of 2×10^{-4}/ml with impulse frequency of 25/second having impulse duration of 20 ms and pause duration of 20 ms. The anodic reaction decreased with rise of impulse frequency. With the increase in conductivity of water to 10×10^{-4}/ml, *I. melanotus* exhibited greater anodic reaction in impulse frequency of 25/second.

The anodic effect on fish was obtained with impulse current of sudden increase and gradual decrease. The duration of each shock was 2 ms. Small fishes gave best results in 20 shocks/second; while larger fish moved to positive pole with 2 shocks/second (Kreutzer and Peglow, 1949) Eighty four per cent of *Gasterosteus aculeatus* showed anodic galvanotaxis and galvanonarcosis with impulse frequency of 65/second in water conductivity of 2×10^{-4}/ml and the anodic effect decreased with the increase of impulse frequency.

The optimum anodic effect of *Tinca tinca* of 191-240 mm during galvanotacis and galvanonarcosis was observed in impulse frequency of 78/second irrespective of conductivity of water in impulse DC of square wave form. The anodic reaction decreased with the rise of impulse frequency.

The optimum impulse frequency for anodic reaction of *Salmo fario* of 81-90 mm was found to be 73/second with impulse duration of 7.2 ms and pause duration of

6.4 ms in water having conductivity of 2×10^{-4}/ml, when 80 per cent of fish moved to anode during galvanotaxis and 100 per cent during galvanonarcosis. The anodic reaction of the fish was reduced in other impulse frequencies. In higher conductivity of 10×10^{-4}/ml also 100 per cent of the fish showed anodic galvanotaxis and narcosis in impulse frequency of 73/second, when impulse DC of square wave form was used.

Narcotizing pulse threshold values were 80, 50, 30, 100 and 40 per second for *S. irideus, C. carpio, I. melanotus, G. aculeatus* and *T. tinca* respectively. As reported by other observers. Biswas (1971) found the narcotizing pulse thresholds for *S. irideus, C. carpio, G. aculeatus, T. tinca* and *S. fario* to be 82, 78, 65, 78, and 73 per second respectively in water having conductivity of 2×10^{-4}/ml. In higher conductivity of water the value of narcotizing pulse threshold remained unchanged. With impulse DC of rectangular wave form the narcolizing pulse threshold for *S. irideus, C. carpio* and *I. melanotus* were 34, 25, and 25 per second respectively irrespective of conductivity of water. Though the narcotizing pulse thresholds for *S. irideus* and *C. carpio* were different in impulse DC of squate wave form and rectangular wave form, the percentage of anodic effect of fishes during galvanotaxis and galvanonarcosis were more in impulse DC of square wave form.

Though the narcotizing pulse threshold for anodic reaction of *S. fario* was found to be 73 per second with impulse duration of 7.2 ms and pause duration of 6.4 ms, yet 100 per cent of fishes were not found to react anodically during galvanotaxis. 20 per cent of fishes moved to cathode during galvanotaxis. Fishes exhibited 100 per cent anodic reaction during galvanotaxis and galvanonarcosis when they lay parallel to current direction with heads towards the anode. 80 per cent of fishes moved towards the cathode when their heads were pointed towards the cathode. The transversal escape movement (Oscillotaxis) occurred when the fish body was perpendicular to the flow of current during galvanotaxis.

Salmo fario of 131-175 mm in an impulse frequency of 73 per second of square wave form showed 100 per cent anodic galvanotaxis when their body axis lay parallel to direction of current with heads towards the anode. When heading towards cathode 50 per cent of fishes moved to anode and 50 per cent to the cathode. Varying position of body axis to field lines from 30 to 70 degree, 100 per cent of fishes reacted anodically. Only when the fish body was perpendicular to current direction, 75 per cent of fishes moved to anode and 20 per cent to cathode. Five per cent of fishes exhibited transverse escape movement.

Studies were made with beheaded *S. irideus* of various size groups in electrical field using an impulse DC of square wave form having frequency of 82/second and impulse duration of 5 ms and pause duration of 7.2 ms. With the rise of threshold current density, 100 per cent of the decapitated fishes moved towards the anode with vibration of the trunk as long as the trunk had power to move, when they were placed in parallel or 45 degree angle to the lines of current conduction of the electrical field. When the beheaded trunks were placed at right angles to current direction, feeble unspecific escape movements were observed. (Biswas, 1971).

Using square wave impulse at frequencies of 76 to 88 per second with variable impulse and pause durations of 4.8 ms to 1.8 seconds for distinct reactions in *Salmo irideus, Cyprinus carpio, Tinca tinca, Gasterosteus aculeatus* and *Salmo fario*, fishes of all the species exhibited more or less similar reactions with the rise of threshold values of current intensity in the field. Besides, sensing the current (first reaction), their movement in the field, with respose to increasing intensity largely depend on their initial position in the field (parallel to current conduction, pointing its head to anode or cathode, or at angle to field lines till 90 degree to lines of current) some times moving violently between electrodes for 15 to 82 seconds and finally stopped near either of the electrodes or in the middle at right angles to the direction of current flow. Matured male *S. irideus* (180-190 mm) ejected milt on bending their bodies at a current density of 0.021 in a voltage tension of 0.502 when subjected to an impulse frequency of 82 per second at a water temperature of 12 degree Celsius and electrical conductivity of 1.6×10^{-4}/ml.

With superiority of impulse frequencies between 82 to 61 per second of varying impulse duration (5.0 to 8.0 ms). no specific conclusion could be drawn with regard to directional swimming of *Salmo fario* (81-90 mm) and *Gasterosteus aculeatus* (36-54 mm) and the threshold values for bringing out those reactions.

Chapter 6

Studies on Warm Water Fishes in DC and Pulsed DC Field

DC Field

Ten varieties of Indian fresh water fishes, namely, *Labeo rohita, Catla catla, Cirrhina mrigala, Labeo fimbriatus, Notopterus notopterus, Channa punctatus, Wallago attu, Mystus aor, Carassius auratus* and *Scatophagus argus* and one variety of freshwater prawn namely, *Macrobrachium rosenbergii*, selected for their anatomical and behavioral peculiarities were subjected to electric shocks to identify and describe responses of animals in relation to field intensity and nature of current.

Behavioral Changes in a Homogeneous Field, Classification of Reactions and Reaction Threshold

Freshwater prawn (*Macrobrachium rosenbergii*) in normal conditions have been found to crawl at the bottom slowly by scanning their surroundings with the help of their antennae. They responded to other external stimuli with a fright reflex by crawling and jumping backwards from the area of disturbance. Sudden illumination momentarily stupefied them but afterwards they moved away from the illuminated area.

On exposing the prawns to under-water DC field of increasing field strength, *Macrobrachium rosenbergii* of 120 to 170 mm length and weighing 25 to 55 gram exhibited reactions in five stages on increasing the intensity in a continuous DC field.

☆ Stage – I Straightening of antenna, backward crawling movement, responded to other external stimuli in a lower degree.

☆ Stage – II Expansion of walking legs, pleopods and straightening of body and tail, placed itself either perpendicular or at a 45 degree angle to the bottom plane. Do not respond to other external stimul including illumination.

☆ Stage – III Curving of body in dorsal position with extension of walking legs and pleopods.

☆ Stage – IV Incapable of any movement, lied on its back with extended walking legs pleopods at the same original place.

☆ Stage – V Jumped out of the field, undergone muscular rigidity, fall on its back at the bottom, dislodging of walking legs from the body (ecdysis).

They required current densities of 0.079 to 1.98 µA/sq.mm for undergoing the above reactions in 100 per cent cases except for the reaction in stage III where the response was just 85 per cent.

The Indian carps, namely, *Labeo rohita, Catla catla* and *Cirrhinus mrigala* exhibited coordinated horizontal swimming movement at the rate of 4 to 6 cm. per second. At times, *Labeo rohita* and *Cirrhinus mrigala* settled at the bottom without any swimming movement. They reacted to other external stimuli sharply, and quickly by moving away from the source of disturbance. Even a shadow on the water disturbed them and the fish quickly moved away from the affected area.

The behavior of *Labeo rohita* in a DC field have been broadly classified in to three groups.

☆ Stage – I: Felt the current with jerks and moved to negative electrode (45.4 per cent). 21.6 per cent of fishes felt the current with jerks and moved to positive electrode. 7.2 per cent fishes felt the current with jerks and placed its body axis perpendicular to field lines.Rest of the fishes (25.8 per cent) felt the current with jerks and slowly moved between the electrodes.

☆ Stage – II: 75.7 per cent fishes moved violently towards the +ve electrode in an elipitical path with a speed of 18-26 cm/second; while 18.7 per cent fishes moved violently towards the –ve electrode with a speed of 12-16 cm/second. Violent movement between electrodes with occasional jumping out of the field placing its body axis pereendicular to field lines were noticed in 5.6 per cent cases.

☆ Stage – III: Immobilization near positive electrode was noticed in 42 per cent fishes. In the center of the field 50 per cent fishes were immobilized. Only 8 per cent fishes were immobilized near negative electrode.

Labeo rohita of 88 to 153 mm length exhibited all the above three reactions in current densities of 0.04 to 0.665 µA/sq mm one after another.

Catla catla of 183 to 270 mm, in a DC field showed the reactions in three stages with increasing current intensity of 0.013 to 0.089 µA/mm and behaved in the following manner.

☆ Stage – I: Stretching of fins, widening of mouth, increased gill movement, tremor of the body, occasionally settled with jerks of body was noticed in all the fishes exposed to current.

☆ Stage - II: All the fishes moved towards the positive electrode.

☆ Stage – III: All the fishes lied unconscious with belly up, extended mouth and fins at the bottom and no longer able to move from that place.

Cirrhinus mrigala, however, in an increasing DC field behaved in a different manner. Fishes of 71 to 105 mm also exhibited the three major reactions in increasing current intensities from 0.008 to 0.16 µA/sq mm.In stage-I, 70 per cent of the fishes showed affinity for the anodic reactions, while the remaining 30 per cent inclined towards the cathodic reactions. In stage – II, 55.8 per cent of the test fishes moved towards the anode, 27.9 per cent moved towards the cathode and 16.3 per cent had undergone curvature of the body extremities pointing towards the positive electrode. 100 per cent fishes were unconscious facing the anode during third stage of reaction.

Featherback, *Notopterus notopterus* have been found to swim horizontally in a coordinated way by slow undulation of the posterior part of the body under normal condition. They maintained their equilibrium by rapid pectoral beats along with undulation of their caudal fin. They were also susceptible to other external stimuli and exhibited fright reflex when the occasion arose.

Notopterus notopterus of 94 to 167 mm, when treated in DC field, their reactions were observed in three broad groups with an increasing current densities of 0.018 to 0.45 µA/sq mm.

☆ Stage – I: 55.2 per cent fishes felt the current with jerks of body and halted. 12.8 per cent fishes felt the current with jerks and moved slowly to +ve electrode. 19.2 per cent fishes felt the current with jerks and moved slowly to –ve electrode. 12.8 per cent fishes moved backward pointing its head to +ve electrode.

☆ Stage – II: 75.5 per cent fishes moved violently towards the positive electrode; while 18 per cent fishes exhibited violent movement towards the –ve electrode and 6.5 per cent fishes moved violently between the electrodes.

☆ Stage – III: Immobility of fishes have been observed near the +ve electrode (57.6 per cent); near the –ve electrode (25.6 per cent) and in the center of the field (16.8 per cent).

Channa punctatus under normal conditions rested at the bottom with a slow pectoral beat and came to water surface at intervals for gulping in air. They were very much less responsive to external stimuli and only moved from their original place, when disturbed physically.

In a DC field, where the field intensity was raised gradually *Channa punctatus* exhibited the following reactions in current densities between0.063 to 0.63 µA/sq mm.

☆ Stage – I: Fifty per cent of the fishes perceived the surrounding electrical field with expansion of dorsal and ventral fins accompanied jerks of the body; while the rest 50 per cent perceived the electrical field and placed its body axis perpendicular to field lines.

☆ Stage – II: All the fishes moved violently towards the positive electrode.

☆ Stage – III: Immobility of fishes were observed near positive electrode (50 per cent) and in the center of the field (50 per cent).

The size range of fishes tested were 85 to 140 mm.

Wallago attu, the scale less catfish, under normal conditions exhibited similar behavior as that of *Notopterus notopterus*.

When exposed to DC field, fishes of 217 to 262 mm *Wallago attu* exhibited reactions in three stages with increasing current density from 0.027 to 0.467 µA/sq mm.

☆ Stage – I: Fifty per cent of fishes moved towards the negative electrode with jerks of the body. 25 per cent fishes moved towards the positive electrode with jerks of the body. Rest 25 per cent placed their body axis perpendicular to field line.

☆ Stage – II: Seventy five per cent fishes moved violently towards the positive electrode with jumps.; while the rest 25 per cent moved violently towards the negative electrode with short undulations of body.

☆ Stage – III: In the third stage, immobility of fishes have been observed near +ve electrode (50 per cent) and near –ve electrode (50 per cent).

Mystus aor, devoid of scales on their body, normally remained at the bottom. They moved to safer place when disturbed by any external stimuli scanning the surroundings with their maxillary barbells. The degree of their response, however, was lesser than that seen in *Notopterus notopterus* and *Wallago attu*.

Mystus aor of 92 to 122 mm exhibited response in four stages in a DC field of increasing intensity of 0.013-0.239 uA/sq.mm. 16.6 per cent of fishes felt the current with jerks of body and moved to either +ve or negative electrode. But the remaining 66.8 per cent, though felt the current with jerks of body, but changed their position perpendicular to current lines. In the second stage all the fishes moved violently to the negative electrode in an elliptical path with tremor of the body. In the third stage, with increasing field intensity all the fishes lost all movements and stayed at the bottom with stretched barbells, expanded fins, jaws and operculum near the negative electrode, pointing their heads towards the –ve electrode. All the fishes succumbed to death in that position.

Carassius auratus have been found to swim in a coordinated manner at a speed of 5 to 6 cm per second under normal condition. A fright reflex accompanied with jerks of the body was noticed when they encountered any external stimulus.

Fishes of 110 to 116 mm length felt the current with expansion of fins, operculum and change of position with jerks of body (68 per cent); slow movement towards the positive electrode with expanded fins (24 per cent) and lying perpendicular to field lines (8 per cent). In the second stage, 36 per cent of fishes either placed themselves at the center of the field or near the negative electrode with expanded fins and operculum perpendicular to current lines. The remaining 28 per cent placed themselves near the +ve electrode in the same condition. Narcosis set up in all the fishes in higher current density at the third stage at different position (20 per cent pointing their head to anode; 32 per cent near the –ve electrode, perpendicular to current lines; 12 per cent in the center of the field and 36 per cent pointing head to –ve electrode with expanded fins and operculum). Current density of 2.3 to 7.4 μA/sq mm were required to set out all these reactions.

Scatophagus argus, acclimatized to freshwater, behaved in a similar way as that of *Notopterus notopterus* under normal conditions. They were found to respond in the DC field in three stages with increasing field intensity. Fishes of 55 to 109 mm length felt current with expansion of fins and jerks of body and remained perpendicular to field lines (75 per cent) and moved slowly towards the +ve electrode (25 per cent). Violent movement with stretched fins and body tremor towards +ve electrode (60 per cent) and to –ve electrode (20 per cent) was observed with rest 20 per cent moving violently in between electrodes. Narcosis set in in all fishes and remained on its side with expanded fins either to positive electrode (50 per cent), or to –ve electrode (15 per cent). 35 per cent fishes were floating on the water surface with ventral portion of the body upward. All these reactions were effected at current densities between 1.8-7.0 μA/sq mm.

Pulsed DC Field

With the rise in field intensity, *Macrobrachium rosenbergii* responded for reactions in five stages in DC current, but in pulsed DC the organisms exhibited reactions in four stages in much lower current densities (0.0198 to 0.594 μA/sq mm in pulsed DC as against 0.079 to 1.98 μA/sq mm in DC field). Elevation of their bodies on the walking legs, backward jumps and to and fro movement of the pleopods at each pulsed during the second and third stages of reactions were noticed in pulsed DC only. With the increase of pulse frequency from 3/second to 7/second, the optimal current densities for the reactions decreased to 0.0198 to 0.4118 μA/sq mm as against 0.0632 to 0.594 μA/sq mm in 3 pulses per second.

Comparing the threshold values for four reactions (first reaction, electrotaxis, electronarcosis and tetanus) of *Labeo rohita* of 45 to 55 mm length, fishes have been found to respond in lower values of 0.0158 to 1.584 μA/sq mm in impulse DC of 3 to 7 per second as against 0.0297 to 1.5998 μA/sq mm in continuous DC field.

Ninety per cent of fishes moved towards the anode during electrotaxis in pulsed DC as against 50 per cent in DC field. Also 100 per cent of them were immobilized near the positive electrode at pulsed DC whereas in DC field 50 per cent of them stayed near the anode and 50 per cent near the cathode during narcosis.

The reaction thresholds have been found to decrease in pulse frequency of 7/second (0.0158 to 1.1959 μA/sq mm) than 0.0317 to 1.584 μA/sq mm in pulse frequency of 3/second. When the fishes were subjected to low frequency pulses (2.5 to 22.2/minute) with variable "on fraction" and "off fraction" at a fixed field intensity of 0.1584 μA/sq mm, *Labeo rohita* of 170 to 310 mm showed 100 per cent anodic taxis and narcosis within 3.5 to 8.5 seconds in a frequency of 22.2 per minute having an "on fraction" of 2.57 seconds and "off fraction" of 0.13 seconds. At frequencies 7.5 per minute, having "on fraction" of 7.73 seconds and "off fraction" of 0.27 seconds and at frequencies 2.5 per minute, having "on fraction" of 23.197 seconds and "off fraction" of 0.803 seconds, 83.4 per cent and 66.8 per cent anodic taxis and narcosis have been observed within 7.5 to 18.6 and 8.5 to 24.8 seconds respectively. On interchanging the duration of "on fraction" and "off fraction", the fishes failed to show any anodic taxis and narcosis except in case of impulse frequency of 2.5 per minute, where 33.2 per cent fishes showed cathodic taxis in 143 to 222 seconds. Rest of the fishes moved either aimlessly between the electrodes or remained perpendicular to field lines with expanded fin without being unconscious set in till 341.5 seconds, irrespective of impulse frequencies.

All the *Cirrhinus mrigala* of 45 to 55 mm length in pulsed DC (3 to 7 pulses per second) exhibited circular movement near the anode during taxis and were narcotized in the same position without showing any movement after switching on the current at a lower threshold values (0.0277 to 1.98 μA/sq mm) than in DC (0.0297 to 0.697 μA/sq mm) where 54 per cent and 14 per cent of them showd movement towards the anode and the cathode respectively and the rest 32 per cent exhibited circular movement near the negative electrode and remained perpendicular to field lines near cathode.

The threshold values for different reactions were decreased in higher frequencies (7 pulsed per second) to 0.0277 to 0.603 μA/sq mm from 0.18 to 1.98 μA/sq mm when treated in pulsed DC of 3 pulses/second. *Cirrhinus mrigala* of 203 to 240 mm length at a peak pulse intensity of 0.1584 μA/sq mm behaved differently in varying pulse "on fraction", frequency being the same (22.2 per minute). 83.4 per cent of the fishes responded for anodic taxis and narcosis within 3.8 to 4.3 seconds when the pulse "on fraction" and "off fraction " were 2.57 and 0.13 seconds respectively. Only 16.6 per cent did not exhibit narcosis, and showed movements between electrodes. On changing the pulse "on fraction" and "off fraction" to 0.13 and 2.57 seconds respectively, all the fishes failed to show specific movement and narcosis, and remained at 45 degree angle to field lines pointing their heads to positive electrode. Body jerks were observed during each pulse.

The superiority of pulsed DC over continuous DC was once again established with respect to threshold values of current densities initiating the different reactions and also of directional swimming and specific movement. *Notopterus notopterus* of 120 to 140 mm length reacted in lower threshold values of 0.0099 to 1.584 μA/sq mm in pulsed DC (3 to 7 pulses per second) than 0.0198 to 0.396 μA/sq mm in continuous DC field. But in pulsed DC, during the second reaction, only 14 per cent of feather backs exhibited forced swimming towards the cathode, while the remaining 86 per cent stayed motionless placing their body axis perpendicular to field lines. In higher intensities, however, all the fishes moved violently towards the anode prior to narcosis. As in many other fishes (*L. rohita, C. mrigala*), the higher pulse frequency (7/second) reduced the threshold values for the reactions still further (0.0099 to 0.307 μA/sq mm) from 0.0693 to 1.584 μA/sq mm required to initiate the reactions pi pulsed DC of 3 pulses per second.

Effect of DC and Pulsed DC on Hybrid Carps (*Catla catla* and *Labeo rohita*)

DC Field

The hybrid fishes in DC field exhibited the perception of electric current in peak body voltage between 0.06 to 0.096 volts, by backward movement, jerks of body and jumping out of water and remained perpendicular to field lines.

With the rise of body voltage between 0.096 to 0.228 volts, they exhibited forced movement between electrodes (50 per cent); violent movement towards positive electrode (25 per cent) and maneuvered to remain perpendicular to field lines (25 per cent) in the second stage of reaction.

In the third stage, at peak body voltage of 0.24 to 0.324, seventy five per cent of fishes were narcotized near positive electrode with head pointing to positive electrode. Rest 25 per cent of fishes undergone narcosis in the mid-field pointing their heads to the anode.

Cramping of the body occurs in 25 per cent of fishes at higher body voltages of 0.324 to 0.54 volts. No death have been observed in DC field at this field strength.

Impulse DC

When the hybrid fishes were exposed to pulsed DC, with pulse frequency of 2, 3 and 5 per second, 100 per cent of them felt the current with expanded fins, and jerk of body at each pulse, when subjected to body voltages of 0.03 to 0.1 volts and pulse frequency of 2 per second. In the higher impulse frequency (5/sec) fishes felt the current with expanded fins and tremor of the body in 75 per cent of the cases at body voltage of 0.012 to 0.024 and with jerks of body in addition to the tremor of the body at 0.024 to 0.036 volts in 25 per cent cases.

Forced movement between electrodes and maneuvered to move to positive electrode with each pulse was observed in all the fishes at 2 pulses/second, body voltage being 0.24 to 0.75 volts. In 3 pulses per second 25 per cent fishes jumped out of water and remained near negative electrode at body voltage between 0.1 to 0.112 volts; while the 25 per cent fishes showed forced movement between electrodes, but ultimately remained near positive electrode. At body voltages of 0.24 to 0.27 volts the rest 50 per cent fishes exhibited forced movement towards the anode at body voltages of 0.12 to 0.24 volts. Violent movements between electrodes, jumping out of water was noticed in 75 per cent fishes with pulse frequency of 5/second at 0.06 to 0.084 volts (body voltage). In still higher body voltage (0.12 to 0.144), 25 per cent fishes after violent movement between electrodes stayed near negative electrode.

Narcosis near positive electrode set in in 80 per cent fishes at 0.36 to 0.9 volts (body voltage) in pulse frequency of 2/second. Hypnosis occurred in 20 per cent fishes near negative electrode at body voltage of 0.36 to 0.66 in the same pulse frequency. In impulse frequency of 3/second 75 and 25 per cent fishes exhibited narcosis near positive electrode and in the mid-field with head pointing towards the anode at body voltages of 0.3 to 0.45 volts. In higher frequency of 5 pulses per second, narcosis set in near negative electrode at higher body voltage of 0.204 to 0.215 volts in 25 per cent cases; while 75 per cent fishes exhibited narcosis near positive electrode at lower body voltage of 0.096 to 0.132 volts.

Death of the animals occurred in 20 per cent cases at two pulses per second at high body voltage (0.9 to 0.95 volts).

In all the current forms (DC and pulsed DC) all the hybrid fishes perceived current at different body voltages, the lowest being 0.012 to 0.024 (5 pulses/second) and the highest in continuous DC (0.036 to 0.096). Anodic galvanotaxis occurred in 50 per cent cases in continuous DC and in 75 per cent cases in pulsed DC of 3/second. Anodic narcosis set in in 100 per cent fishes in pulsed DC of 2/second followed by 75 per cent each in continuous DC and pulsed DC of 3 pulses/second. In 5 pulses per second, half of the fishes showed anodic narcosis, while rest half showed cathodic hypnosis.

Chapter 7

Studies on Warm Water Fishes in AC Field (Freshwater Fishes)

The alternating current (AC) is characterized by a sequence of positive and negative waves which are equal, usually sinusoidal and which follow each other alternately at regular time intervals. Electric fishes among Torpedids, Gymnotids make use of this type of cyclic current generated from their electric organs for catching their prey. Gymnotids are characterized by the discharge of regular monophasic pulses, which vary in frequency (number of cycles per unit time) from 60 to 400 discharges per second (Olson, 1972). Many authors were therefore keenly interested in the studies of the response of fish to AC field.

Sharks and Rays were reported to be extremely sensitive to AC field (Kalmijn, 1966).

Shetter (1947), was of the opinion that, the rapid reversal of electric current flow in an AC circuit affects the nervous system of fish.

Meyer-Waarden (1957), while describing the reaction of fish in alternating current stated that the fish did not swim towards one of the two electrodes, as they do in the case of direct and interrupted direct current, but they took a transverse position to the direction of the current, between the two electrodes, in such a way that they would tap off a minimum voltage (Oscillotaxis).

According to Scheminzky *et al* (1941) most of the fish (for instance, trout, carp, tench, catfish etc.), undergo a kind of "hypnosis" after the current is switched off. They do not return immediately to normal swimming position, but stay for some minutes in a lateral or dorsal position. When the "hypnosis" wears off, the fishes swim away as they do in the case of direct current.

Haskell (1950) reported the use of AC electric shocker for collection of fish in 1940. Pratt (1952) carried out tests on the efficiency of alternating current fish shocker over that of the direct current ones. McLean *et al* (1953) described the development of AC devices as a most promising means of controlling the parasitic sea lamprey in the Great Lakes of USA. Kuriki (1953) reported the successful use of alternating current of sinusoidal wave, 60 cycles, in the electric fish screen. The electric field activated by 110 volts, 60 cycles alternating current has been stated to have worked satisfactorily in diverting a sufficient number of adult Chinook salmon into the spawning ponds (Applegate *et al*, 1962).

Knowledge of the conditions of the action of direct current, according to whether the fish faces the anode or the cathode, enabled a study of alternating sinusoidal carrent, where the fish successively faces the anode and the cathode as many times per second as the frequency dictates; and accordingly comparative studies have been made on the reactions of fish to alternating sinusoidal currents by Halsband *et al* (1960), Lamarque (1963) and Blanchetau (1961). The behavior, physiological changes and electro-sensitivity of different fish species to alternating current were also studied by Bodrova *et al* (1958, 1059), Lukashov *et al* (1963) and Shentiakov (1959).

All these studies were on fishes other than Indian species. Hence the tests on three Indian freshwater fish species, namely, *Puntius ticto, Heteropneustis fossilis* and *Tilapia mossambica* have been carried out for their electro-sensitivity and response to AC field (Biswas, 1974).

The current form used for these studies was sinusoidal wave having a frequency of 50 Hz. per second available from the mains supply.

Behavioral Changes in a Homogeneous AC Field

Identifiable reactions in successive stages occurred in fishes with the increasing field strength, could be classified as below.

Stage – I (First Reaction)

Puntius ticto could perceive the surrounding electrical field by stretching their dorsal, pectoral, ventral and anal fins. Occasional jerks of the body being parallel to the lines of current conduction was often followed by tremor of the posterior part of the body with the increasing field strength. The opercular beat slowed down and the fish did not react sharply to external stimuli. Slow coordinated movements between the electrodes continued in this stage.

Heteropneustis fossilis in this stage exhibited jerks of the head on being parallel to current lines. Bending of the posterior part of the body with tremor of the caudal fin was noticed in higher current intensity. Response to external stimuli was reduced to a considerable extent, with a slowing of gill movement.

Tilapia mossambica could feel the presence of electrical field and showed 8 to 42 per cent increase in the rate of opercular movement. It exhibited jerks of the body when it was parallel to current lines. Normal coordinated swimming slowed down and the fish did not react to the external stimuli

Stage – II (Longitudinal Oscillotaxis)

Jerky swimming of *Puntius ticto* between the electrodes, with vibrations of the posterior part of the body was initiated. Violent contraction of the body forming a bend along the body axis resulted unbalanced swimming in an elliptical path. A few fishes jumped out of the field and floated on the water surface.

In *Heteropneustis fossilis* forced swimming with strong vibration of the body set in, without any coordinated movement. The fish moved between the electrodes making an angle of 45 degree with their body axis to the current lines. They also swam between the surface and the bottom of water media making 45 to 65 degree angle with their body axis to the electrode plane. Occasionally individuals moved between the electrodes lifting the posterior part of the body at an angle of 45 degree to the bottom plane.

Involuntary movement between the electrodes with expanded fins, and bending of body started in *Tilapia mossambica* in this stage. Strong vibrations of the body accompanied the unbalanced movement and the fish took an elliptical path.

None of the above mentioned species responded to any external stimuli at this stage.

Stage – III (Transverse Oscillotaxis)

At a higher field intensity, *Puntius ticto* finally placed itself perpendicular to the field lines and at times with bottom plane without any further movement. Rate of gill movement slowed down further and showed irregular beats. When shifted to any other position with the help of a glass rod, the fish oriented itself to its original perpendicular position in the field.

Heteropneustis fossilis and *Tilapia mossambica* also adopted a static position, at a 90 degree angle to the current lines, in the similar way as *Puntius ticto* described earlier. However in the case of *Heteropneustis fossilis* the black body color faded to a light yellow at this stage; while in *Tilapia mossambica* the grayish body color darkened to a deep grayish black with reddish tinge on the margins of the fins.

Stage – IV (Tetanus)

Muscular rigidity, in *Puntius ticto* was accompanied by the full stretching out of the dorsal, pectoral, ventral and anal fins and by the tetanus of maxilla and gill cover, the latter resulting in the closure of the operculum. The fish either settled down on its side or remained upright at the bottom with no visible sign of body movement. Occasionally some individuals floated on the surface of water with their belly up. The pupil were found to be dialated.

Tetanus of *Heteropneustis fossilis* set in with the contraction of maxilla and gill cover in a fully stretched condition and the bending of the body like an arch. All voluntary movements stopped and the fish either sank to the bottom on its side or remained their head standing erect resting on the posterior part of the body, perpendicular to the bottom plane. Dorsal, pectoral and ventral fins remained fully stretched in this stage. Alternate contractions and relaxations of myomeres were observed under this condition. The body color from light yellow deepened to a deep yellow. Dialation of the pupils were also noticed.

The titanic condition of *Tilapia mossambica* set in with the stoppage of all voluntary movements and sinking of the fish to the bottom on its side with stretched dorsal, ventral and anal fins. Contraction of maxilla and the closure of gill covers and dialation of pupil also occurred at this stage. The grayish black body color faded to white when facing the electrode. On being perpendicular to current lines, the color of the fish darkened to deep black. Spasmic ejaculation of milt from males continued under the tetanized condition.

Stage – V (Hypnosis)

After the stoppage of current flow the entire body relaxed from the tetanized state and the gill cover started beating slowly in a regular rhythm. The rate of gill movements gradually increased and regained 52 to 95 per cent of the normal rate in *Puntius ticto*. All the fins remained in a stretched condition with no sign of voluntary movements. The fish remained in this condition for 21 to 274 seconds.

Heteropneustis fossilis also continued in a hypnotic state immediately after the stoppage of current flow for 1 to 2 seconds. During this condition the fish did not exhibit any voluntary movement. Operculum started beating slowly at the beginning and regained its normal rate by 62 to 93 per cent after the vanishing of the hypnotic state.

Tilapia mossambica remained under the hypnotic condition with expanded dorsal, ventral and anal fins for a period of 5 to 165 seconds after the switching off the current. During this condition the fish did not exhibit any movement except that of the gill cover which resumed slowly.

Behavioral Changes during Recovery from Hypnosis

Followed by hypnosis, all the species under study returned to normal swimming position. Pectorals started beating and the fish manoeuvred to prevent rolling on its side. 10 to 12 per cent increase in gill movement from the normal, was noticed in *Puntius ticto*; while 9 to 26 per cent rise of gill movement was observed in *Heteropneustis fossilis*. But 6 to 28 per cent decrease from normal rate of gill movement was recorded in the case of *Tilapia mossambica* after the recovery. *Heteropneustis fossilis* regained its original black body color within 128 to 144 seconds and *Tilapia mossambica* regained its original grayish color within 20 to 40 seconds after recovery.

Occurrence of Identifiable Reactions

The presence of identifiable reactions as classified earlier, occurred in fishes of the above three species. Stretching of dorsal, ventral, pectoral and anal fins were observed in 27 to 62 per cent of fishes in the case of *Puntius ticto*, 48 to 74 per cent in the case of *Heteropneustis fossilis* and 37 to 54 per cent in the case of *Tilapia mossambica*. Jerks of body occurred in 0 to 85 per cent, 0 to 68 per cent and 0 to 47 per cent in *Puntius ticto, Heteropneustis fossilis* and *Tilapia mossambica* respectively, when they were parallel to the lines of current conduction. Quivering of posterior part of the body was noticed in 16 to 53 per cent, 5 to 56 per cent and 2 to 23 per cent of the above mentioned species respectively. Jerky swimming between electrodes was observed in 0 to 84 per cent of *Puntius ticto,* (fishes when parallel to current lines.) to 65 per cent of *Heteropneustis fossilis* and *Tilapia mossambica* had undergone forced swimming under similar condition. Transverse oscillotaxis set in, in 47 to 100 per cent of all the above varieties of fish when perpendicular to current lines. Muscular rigidity accompanied by tetanus of gill cover and stoppage of opercular movement took place in all the fishes irrespectivie of varieties. Alternate contractions and relaxations of myomeres during titanic condition were observed in *Heteropneustis fossilis* only. Spasmic ejaculation of milt from male *Tilapia mossambica* was recorded in 0 to 64 per cent of the test fishes. A change of body color during current exposure took place in 100 per cent of *Heteropneustis fossis* and *Tilapia mossambica.* 18 to 62 per cent of *Puntius ticto,* 39 to 76 per cent of *Heteropneustis fossilis* and 12 to 385 of *Tilapia mossambica* remained under hypnotic condition before returning to normal swimming condition even after switching off the current.

Field Intensity and Stages of Reaction

The threshold values of current densities and voltage gradient for different stages of reactions of *Puntius ticto* of 56 to 95 mm; *Heteropneustis fossilis* of 106 to 170 mm and *Tilapia mossambica* of 121 to 170 mm revealed that the first reaction was initiated in current densities of 0.13 to 0.25 µA/sq mm in *Puntius ticto,* 0.06 to 0.3 µA/sq mm in *Heteropneustis fossilis* and 0.045 to 09 µA/sq mm in *Tilapia mossambica.* They exhibitedlgitudinal oscillotaxis in current densities of 0.28 t0 0.66 µA/sq mm, 0.18 to 0.6 µA/sq mm and 0.06 to 0.22 µA/sq mmrespectively. The same species entered into transverse oscillotaxis in current densities 0.6 to 0.7 µA/sq mm, 0.21 to 0.66 µA/sq mm and 0.16 to 0.27 µA/sq mm respectively. The coponding values of current densities for effecting tetanus were, however, 0.64 to 0.98 µA/sq mm, 0.42 to 0.78 µA/sq mm and 0.36 to 0.44 µA/sq mm for *Puntius ticto, Heteropneustis fossilis* and *Tilapia mossambica.*

The ratio of reaction thresholds were calculated to be;

1:2.1:4.6:5 to 1:2.6:2.8:4 for *Puntius ticto;* 1:3:3.5:7 to 1:2:2.2:2.6 for *Heteropneustis fossilis;* and 1:1.3:3.5:8 to 1:2.4:3:5 for *Tilapia mossambica.*

The relation between the voltage gradient and the field intensity was almost constant for the whole study, which depended on the electrical conductivity of the water media. To effect the field intensity of 0.01 µA/sq mm, voltage gradient of 15.99 to 19.2 millivolt per cm were required irrespective of the fish species studied.

Studies on Warm Water Marine Fishes in AC Field

The possibility of using electricity in conjunction with conventional gear in marine fishing has been explored in the past many years. The idea was to attract and stun the fish so that they would tumble helplessly into the trawl net. Kreutzer and Peglow (1949) claimed that a deep sea electrical trawl, using short sharp electrical impulses up to 2000 volts on oppositely charged electrodes placed at the side of the net's mouth, was able to get 90 per cent of the fish passing between the ship and the net, which would otherwise be frightened by the approach of ordinary net and evade it. In uniform electric field of alternating current of 50 Hz, slight specific differences were observed in the voltage gradients to induce minimum response and electro-narcosis in marine fishes (Bary, 1956). Cattley (1955) recommended the use of pulsed AC for reducing the magnitude of power requirement for fishing in sea, as use of continuous AC would have been impracticable due to the difficulty in carrying generating equipment aboard the vessel to produce enough power.

The reactions of marine warm water fishes in a homogeneous electric field of alternating current of 50 Hz in relation to fish length, repeated shocking and fatigue and position of fish body in the electrical field were carried out with respect of seventeen fish species of Indian west coast (Biswas, 1971). The effective periods along with pulse threshold for inducing electrotaxis, narcosis and recovery were also studied.

Interrupted AC of 220 volt was produced through a mechanical impulse transmitter of fixed impulse frequencies of 38 per minute with an impulse and pause durations of 0.8 second and applied to different fish species and their behavior observed for each fish separately.

Fishes were subjected to interrupted AC of 220 volt to form a homogeneous electrical field throughout the water column. Sixty amperes of current was drained with impulse potential of 130-140 volt between the electrodes while closing the circuit. Since the electrical conductivity of the sea water was very high (300-800 mho/ml) as compared with fresh water (9000-18000 mho/ml), gradual increase of current intensity and voltage gradient in the water media could not be made with the fear of being short circuited between the electrodes. The density of current in the experimental bath was expressed as μA/sq mm, which was kept constant for every set of experiment.

Reactions Threshold and Effective Periods

The reaction of fishes in relation to exposure to current for different periods and the requirement of effective period in relation to size, area of body surface and position of fish body in electrical field were studied in different varieties of fish. The effect of repeated stimulation on the fish in relation to fatigue and recovery was also determined. Except in case of repeated shocking, a fresh fish was taken for each test. The area of body surface was measured by a polar planimeter after the test. The potential difference between the head and tail of the fish was also recorded during electronarcosis and fixation.

In a homogeneous field having a density of current of 1.2 mA/sq mm with fixed impulse frequency of 38/minute, the fish exhibited reactions like stretching of fins and opercula and tremor of the body accompanied by occasional change of place, depending on the minimum period for which the current passed through the body of fish. This minimum flow of current through the fish, defined as "effective period" increased with the administration of increased number of pulses. With the rise of effective period to its threshold value the fish underwent electrotaxis accompanied either by specific movement to the electrode or at random in the field. In some cases the fish did not show any movement in the electric field during electrotaxis but changed its position to right angle to the direction of current flow and bent its body like an arch in the same place to tap off the minimum voltage. Electronarcosis and fixation occurred with more impulses for threshold value of effective periods when the fish was stunned ceasing all the voluntary movements and lay on its side at the bottom or floated on the surface (Biswas, 1971).

In a fixed current density of 1.2 mA/sq mm, the effective periods for electronarcosis and fixation of seventeen varieties of marine fishes ranged from 2 to 6.8 seconds, irrespective their positions in the electric field.

The specific movements of fishes in the study have been classified into four groups during electrotaxis and three during electronorcosis (Biswas, 1971). Meyer-Waarden (1955) also reported that the fish did not swim towards the electrodes in an AC field, but took a transverse position to the electrical field between the electrodes. Fishes of sixteen varieties out of seventeen tested were found to move towards increasing potential during electrotaxis and narcosis, the percentage of which varied from species to species. Oscillotactic behavior has also been noticed in 14 varieties and 12 to 58 per cent of fishes placed their body axis perpendicular to the electrical lines of force to tap off the minimum voltage. Only eight species exhibited movement towards the counter pole during electrotaxis which varied between 8 and 50 per cent. Fishes like, *Apiniphelus diacanthus, Plutosus arab, Scolopsis leucotaenia, Apogon faciatus, Therapon jarbua* and *Dasyatis gerrardii* showed no movement during electrotaxis and stayed in their initial position till the narcosis set in with increase of effective period.

In a constant field intensity of 1.2 mA/sq mm of interrupted AC of sinus wave form, *Crysophus burda* of 420 mm length required an effective period of 1.6 to 4.8 seconds for narcosis and fixation as against 3.2 to 6.4 seconds in case of *Muraena undulate, Epinephelus pantherinus and Dasyatis gerrardii*. The optimum effective period for narcosis of *Crysophus burda* of 550 mm length was also lowest (1.6 seconds) among the other three species when they were stunned in 2.4 seconds. Slight differences in voltage gradient of AC required to induce minimum response and electronarcosis have been observed between *Morone labrax, Platichthys flesus* and *Mugil auratus* (Bary, 1956). *Megalops cypernoides* of 360 mm were narcotized in between 1.6 and 2.4 seconds; while *Lates calcarifer* and *Dasyatis gerrardii* were stunned later successively in 3.2 to 4 seconds and 4.8 to 6.4 seconds respectively. Electronarcosis and fixation were induced in *Tetradon* sp of 100 mm length in 5.6 to 8 seconds, while *Haetodon ollaris, Platax teira, Therapon jarbua, Scolopsis leucotaenia* and *Apogon faciatus* of similar size exhibited hypnosis in 3.2 to 4.8; 2.4 to 4; 1.6 to 5.6; 1.6 to 4.8 and 1.6 to 4 seconds respectively. *Apogon faciatus* of 95 mm was stunned earlier in 1.6 seconds, while *Scolopsis leucotaenia* and *Tetradon* sp. of same size were narcotized subsequently in 2.4 and 6.4 seconds. respectively.

Body Voltage and Effective Period for Inducing Narcosis in an Interrupted AC

At a fixed impulse frequency of 38 per minute and current density of 1.2mA/sq mm, *Apiniphelus diacanthus* of 142 to 180 mm length exhibited narcosis in body voltage of 7.87 to 12.6 in 2.4 to 4 seconds. *Plutosus arab* of 215 to 312 mm were stunned in 1.6 to 3.2 seconds when the voltage drop between their head and tail were 12.9 to 18.7.

Chiloscyllium indicum between 400 and 450 mm length were hypnotized in 2.4 to 5.6 seconds when subjected to body voltages of 23.7 to 27 volts. Potential drops of 21 to 30 volts brought about narcosis and fixation of *Muraena undulate* of 400 to 415 mm in 2.4 to 6.4 seconds. *Scolopsis leucotaenia* with lengths of 95 to 100 mm required a potential drops of 2.3 to 5.7 for paralysis in 1.6 to 4.8 seconds. *Tetradons* of 95 to 300 mm were narcotized in 4.8 to 7.2 seconds when subjected to body voltages of 0.6 to 1.8. The body voltages for narcosis and fixation varied according to length and position of the fish body in the electrical field. The fish which lay parallel to the direction of current conduction was subjected to more potential drop between head and tail than in any angle to the direction of electrical lines or perpendicular to the lines of force for a given size.

Apogon faciatus of 95 to 100 mm, *Platax teira* of 100 to 160 mm, *Siganus vermiculatus* of 260 mm, *Crysophus burda* of 410 to 420 mm, *Lates calcarifer* of 360 to 370 mm, *Megalops cypernoides* of 335 to 350 mm, *Haetodon ollaris* of 33 to 100 mm, *Heniochus macrolepidotus* of 60 mm, *Epinephelus pantherinus* of 550 to 570 mm, *Therapon jarbua* of 100 to 128 mm and *Dasyatis gerrardii* of 385 to 420 mm were narcotized in body voltages of 1.8 to 6, 6 to 9.6, 5.6 to 21.6, 18.4 to 25.2, 16.2 to 22.2, 14.8 to 21, 1.2 to 6, 2.4 to 3.6, 8.1 to 34.2, 1.5 to 7.7 and 11.1 to 25.2 volts respectively.

The effective periods in the present study, for narcosis and fixation of fish were more with the decrease of potential drop between head and tail irrespective of varieties. In *Apiniphelus diacanthus*, the period for narcosis was raised to 4 seconds from 2.4 when the body voltage was lowered to 7.87 from 10. Effective period of narcosis in *Muraena undulate* increased to 5.6 seconds from 4.8 seconds when body voltage was decreased to 2.4 from 24.9. *Scolopsis leucotaenia* of 100 mm required 4.8 seconds for stunning, when subjected to body voltage of 2.3. On increasing the body voltage to 6, the period for narcosis was reduced to 1.6 seconds. Similar reduction in effective periods for narcosis has been noticed in *Siganus vermiculatus* of 260 mm. *Therapon jarbua* of 120 mm and *Dasyatis gerrardii* of 420 mm with increase of potential difference.

Apiniphelus diacanthus of 149 to 180 mm exhibited narcosis and fixation only when the exposure to current was for 2.4 seconds or more in body voltages of 10 to 12.6. A decrease in body voltage to 7.87 increased the effective period for narcosis to 4 seconds. Though the requirement of effective period for narcosis of fish did not vary from 2.4 seconds between 142 to 180 mm of fish length, yet the intensity of shock in large sized fishes was great and could regain senses only in 9.5 seconds in case of 180 mm as against 7 seconds in case of 142 mm.

Plutosus arab between 215 to 330 mm felt the impulse with sudden jerk and bending of body accompanied by tremor. Movement of fishes occurred during electrotaxis and were narcotized in 1.6 to 2.4 seconds in every case at body voltages between 12.9 to 19.8. Fishes of 215 mm required a greater effective period of 2.4 seconds, while those between 312 to 330 mm were narcotized in 1.6 seconds. With the gradual increase in period of exposure to 5.6 seconds, an intense effect on the fish of 330 mm was observed delaying the recovery of fish up to 126 seconds depending on the period of exposure.

The effective periods for initiating taxis and narcosis of *Chiloscyllium indicum* varied inversely from 3.2 to 5.6 seconds with decrease of length from 450 to 400 mm in potential differences of 27 to 24 volts between their heads and tails. Movement of fish accompanied with intermittent jerks during each pulse occurred during electrotaxis. On reaching the threshold effective period, the fish underwent hypnosis. The periods of recovery from hypnosis were longer with the increase of periods of stimulation for the fish of same length.

Marine eel (*Muraena undulate*) of 415 to 500 mm showed narcosis and fixation in 2.4 to 4.8 seconds with gradual increase in effective periods with the decrease of fish length when subjected to potential difference of 24.9 to 30 volts. Violent movements were observed prior to narcosis. The period of recovery of fish was delayed to 3.8 seconds from 1.2 seconds with increase of duration of shocks to 7.2 seconds from 2.4 seconds.

In a fixed density of current of 1.2 mA/sq mm, *Tetradon* sp. of 300 mm remained stunned in 4.8 seconds, while 5.6 seconds were required to narcotize a fish of 100 mm.

Effective periods of 1.6, 3.2 and 4.8 seconds induced narcosis in *Scolopsis leucotaenia* of 100 mm when subjected to voltage tensions of 6, 4.5 and 2.3 volts respectively. Increased periods of shocks intensified the effect on the fish causing delay in recovery in case of both *Scolopsis leucotaenia* and *Apogon faciatus.*

Platax teira of 160 mm were narcotized in 1.6 seconds when subjected to a body voltage of 8.6 volts, while that of 100 mm required 2.4 seconds when the voltage tension was 6.

At potential differences of 21.6, 25.2 and 22.2 volts created between the head and tail of *Siganus vermiculatus* of 260 mm, *Crysophus burda* of 420 mm and *Lates calcarifer* of 370 mm, intense effects of narcosis were observed with the increase in effective periods resulting in delayed recovery of fishes.

Though *Megalops cypernoides* of both 350 mm and 337 mm required 1.6 to 2.4 seconds to exhibit narcosis in body voltages of 21 and 20.2 volts, yet the effects on longer fishes were more prolonged than on the smaller ones as the latter recovered in 6.5 to 8 seconds, while the former took 180 to 252 seconds to regain their normal positions.

Haetodon ollaris of 100 mm were stunned in 3.2 seconds as against 4 seconds required for a fish of 33 mm.

Epinephelus pantherinus of 570 mm, *Heniochus macrolepidotus* of 60 mm and *Haetodon ollaris* of 100 mm took longer periods for recovery with increased periods of stimulation when treated with potential differences of 34.2, 2.4 and 6 volts respectively. Movement during taxis prior to narcosis occurred with expansion of gills, fins and tremor of the body in most of the varieties.

Therapon jarbua of 128 mm, required a higher potential of 7.7 volts than that of 100 mm (6 volts) for narcosis. The effective period for narcosis however remained 1.6 seconds irrespective of variation in length of fish. But the effect on bigger fish was so intense that it took 16 seconds for recovery, while fishes of 120 mm, 105 mm, 103 mm, and 100 mm recovered in 9, 7.5, 4, and 2 seconds respectively. Unlike other marine species, 80 per cent of *Therapon jarbua* did not exhibit any movement during electrotaxis and were narcotized in their originally occupied place in the field with expanded gills and fins. On prolonge exposure to the current for 8.8 to 12 seconds, *Therapon jarbua* of 100 to 128 mm length did not recover from the shock and died within 7 to 20 minutes after the exposure. The periods of narcosis were also prolonged with increase in effective periods and fishes of 128 mm, 120 mm, 103 mm and 100 mm recovered in 61, 17, 14 and 3 seconds respectively.when subjected to electric shocks of 4, 2.4, 3.2 and 3.2 seconds.

Desyatis gerrardii, 285 to 420 mm, felt the electric current by bending of sides accompanied by tremor of body during taxis and were stunned in the same place without any movement in the electric field. Fishes of 420 mm were narcotized in 3.2 seconds, whereas those of 385 mm took 4.8 seconds for narcosis. Intensive effects of shocks have been observed with increase in effective periods irrespective of size of fishes.

Higher voltage gradients were needed to produce the minimum response in a fatigued fish than in a fresh or rested fish in an AC field (Bary, 1956). The effect of interrupted AC on the general behavior of the fish revealed that the rate of gill movement of *Platax teira, Haetadon ollaris, Crysophus burda, Siganus vermiculatus, Scolopsis leucotaenia* increased with treatment of impulse current of different effective periods irrespective of size of fishes indicating increased respiration after shock treatment.

No significant effect was observed on the period for narcosis and recovery in relation to repeated stimulation in case of *Catla catla* and *Cirrhinus mrigala* which was possibly due to the accommodation of nerves and subsequent fatigue of the fish (Biswas, 1969).

Lowered activity was observed in case of all other species tested as shown by decreased rate of gill movements after electric shocks of different effective periods irrespective of fish length.

Summing up, in an interrupted AC of 50 Hz having a fixed impulse frequency of 38 per minute and constant current density of 1.2 mA/sq mm, marine fishes of west coast of India exhibited electrotaxis and narcosis on reaching the optimum period of flow of current through them. The visible reactions were stretching of fins, tremor of the body and change of place accompanied by either specific movement towards the electrodes or at random in the field till the narcosis set in. Some fishes like, *Therapon jarbua* and *Dasyatis gerrardii*, underwent narcosis in the same place without any movement in the electric field. Though the requirements of minimum current flow varied from species to species, it was within the range of 2 to 6.8 seconds irrespective of fish size at a fixed impulse duration of 0.8 seconds. 12.5 to 75 per cent of 16 species moved towards the increasing potential during electrotaxis and narcosis, while 12 to 58 per cent of 14 varieties exhibited oscillotactic reaction and placed their body axis perpendicular to lines of force. Only 8 to 50 per cent of eight varieties swam towards the counter pole during electrotaxis.

Crysophus burda was narcotized in 1.6 to 4.8 seconds, while *Muraena undulate, Epinephelus pantherinus* and *Dasyatis gerrardii* were stunned in 2.4 to 6.4 seconds at potential drops of 11.1 to 33 volts between their heads and tails. Similarly *Megalops cypernoides* reacted earlier (1.6 to 2.4 seconds) than *Lates calcarifer,* and *Dasyatis gerrardii.* *Tetradon* sp also showed narcosis in higher effective period compared to *Haetodon ollaris, Platax teira, Therapon jarbua, Scolopsis leucotaenia* and *Apogon faciatus.*

The effective periods required for narcosis and fixation varied inversely with the potential drop created between head and tail of the fish. Fishes of greater length were subjected to more voltage tension than smaller ones, causing earlier effect on large sized fish in smaller effective period. The same fish which placed itself parallel to the field lines was subjected to higher potential difference between head and tail than that at an angle to electric lines. With increase in angle, the voltage tension between head and tail decreased and was lowest at right angles to the flow of current conduction causing the requirement of higher effective period for narcosis and fixation of fishes. The intensity of electric shock on fishes was more with increase in period of exposure to current and took longer period for recovery. An interrupted AC of 50 Hz affected fishes with a difference in rate of gill movements before and after shock treatment.

Effect of AC Field on Marine Crustraceans

Higman (1956) conducted experimts in a tank containing sea water to determine the optimal electric conditions that would cause maximum movement of shrimp (*Penaeus duorarum*) to the positive electrode in a field of pulsed DC. Biswas (1971) investigated on the behavior of marine shrimps (*Metapenaeus affinis, Parapenaeopsis stylifera,*), lobster (*Panulirus ornatus*) and marine crab (*Neptunus sanguinolentus*) in an AC field in a tank containing sea water, so as to ascertain if AC field will be in a position to force the hidden animals to rise and appear before the catching gear in helplessly stunned condition to pick up.

The study was carried out in sea water (electrical conductivity 750 mho/ml) at a fixed current density of 1.2mA/sq mm at a water temperature of 30 degree Celsius.

Fourteen trials were conducted with *M. affinis* of 65 to 52 mm in an AC field having a current density of 1.2 mA/sq mm. The time required for electronarcosis and fixation depended on the potential difference between the head and tail, which again depended on its initial position to the field lines. Potential difference between the head and tail was maximum when the animal rested with its body axis parallel to the direction of current and minimum when it lay perpendicular to the field lines. All the shrimps, irrespective of sizes, jumped backward and stunned during electronarcosis and fixation. The animal of 55 mm length required 3.2, 4, 4.8 and 8 seconds for narcosis and fixation when the potential drops between their head and tail were 4.1, 3.3, 2.3 and 0.6 volt respectively depending on their position in the electrical field.

Higher potential difference between head and tail induced electronarcosis and fixation of the animal earlier with the increase in length of cephalothorax. Organisms of 58, 56, 55, and 53 mm when placed parallel to the direction of current flow required potential differences of 4.3, 4.2, 4.1 and 3.9 volt respectively between their heads and tails and were narcotized in 1.6, 3, 3.2 and 3.2 seconds respectively.

P. stylifera under similar conditions required higher potential drop between head and tail with increase in length. The effective period for narcosis and fixation varied inversely with the body voltage between the head and tail. The shrimp of 77, 66, 60 and 45 mm length exhibited narcosis and fixation in 1.6, 2.4, 2.4 and 3.2 seconds respectively when the voltages between their heads and tails were 5.8, 4.9, 4.5 and 3.4. Animals, exhibited backward jumping on switching on the current, (all animals), but narcosis occurred in 40 per cent cases irrespective of sizes and their position in the field. Rest of the animals exhibited no narcosis. The voltage tension between head and tail, the effective period (period for which the organism was exposed to the current) was important factor in bringing about electro-narcosis. Complete narcosis occurred in *P. stylifera* of 77 mm length in 1.6 seconds when the potential difference been the head and tail was 5.8 volts. But the organism in similar conditions could not exhibit narcosis when exposed to current for 0.8 seconds.

P. ornatus between 255 and 180 mm showed narcosis with contraction of limbs and tail with switching on the current and were no longer able to move of their own accord as long as the current continued to flow. The animal recovered immediately after switching off the current flow and disappearance of electrical field and exhibited backward movement towards the electrode. Narcosis and fixation of the animal occurred when the period of exposure to electric field reached the threshold value. Animals of 255, 232, 185 and 180 mm were not narcotized when they were exposed to current for 6, 8, 3, 2 and 2 seconds respectively at potential differences of 19.1, 17.4, 16.9, 13.8 and 13.5 volts between their heads and tails. But they were completely narcotized when exposed for 2.4, 3.2, 4.5 and 5.5 seconds respectively under similar voltage tension between head and tail. Forward movement of the organism has been observed after the recovery from narcosis. The effective period required for narcosis of the animal increased with decrease in size. The greater the length, the lesser the time required for narcosis and fixation.

Unlike shrimps and lobster, crab (*N. sanguinolentus*) behaved differently in electrical field of similar conditions. The period required for hypnosis and fixation of the animal did not vary on the size of animal as observed in the case of shrimps and lobster. Crabs of 116 mm carapace length exhibited narcosis in 3 seconds, when crabs of 120 mm and 106 mm were narcotized in 4.2 seconds. Even crabs of similar size (106 mm carapace length) under similar voltage tension of 7.9 volts between extremities of carapace, took 1.8 to 4.2 seconds for their narcosis and fixation. Meyer-Waarden (1954) reported death of all Chinese crabs within 24 hours when they were exposed to impulse current with loss of all their legs in 80 per cent of animals and partly in 20 per cent. Twenty per cent of *N. sanguinolentus* lost their limbs from their joint in the tests.

Using interrupted AC of 50 cycles per second, and impulse voltage of 150 and 37 pulses per minute with impulse and pause duration of 0.8 second, narcosis of shrimps did not occur in 2 pulses having effective period of 1.6 seconds, when subjected to potential drop of 0.6 to 9.6 volts in *M. affinis* of 132 mm to 90 mm. Narcosis and fixation occurred first with 3 pulses of effective period of 2.4 seconds in animals having body voltages between 6.7 to 9.6 volts. With the rise of effective period of 3.2 second in 4 pulses, animals having body voltages 0.6 to 1.9 volts were not affected.

All the organisms were found to be narcotized when treated with 6 pulses having effective period of 4.8 seconds irrespective of the value of potential drops between the head and tail. The animal exhibited electro-narcosis and fixation in every case with the rise of effective period between 4.8 to 6.4 seconds subsequent to the increase of pulses from 6 to 8. In a constant pulse rate of 37 per minute and current density of 1.2 mA/sq mm, six pulses of 4.8 seconds were found to be the optimum, when all the shrimps between 132 to 90 mm were narcotized completely. No significant effect has been observed on the requirement of effective period for the narcosis and fixation of *M. affinis* with repeated stimulation. Prawns between 96 and 91 mm took 1.6 seconds to be narcotized in successive stimulations (Biswas, 1971).

Biswas *et al* (1969) also observed similar effect on *Cirrhinus mrigala*.

Summing up, *M. affinis, P. stylifera* and *P. ornatus* exhibited electronarcosis and fixation in different times when exposed to current density of 1.2 mA/sq mm at temperature and electrical conductivity of 30 degree Celsius and 750 mho/ml respectively, the periods varying with the length of animals.

In case of marine crab, *N. sanguinolentus*, however, the increase in size did not have any effect on the effective periods for bringing about fixation of the animal.

M. affinis required 3 pulses of effective period of 2.4 seconds to bring out narcosis and fixation for length between 132 to 90 mm in interrupted AC of 37 pulses per minute, when the body voltage between head and tail were 9.6 to 6.7. The effective period for narcosis and fixation of the animal varied inversely with the voltage tension between head and tail. The period of recovery of animals increased when treated higher pulses for greater effective periods. Repeated stimulations had no marked effect on *M. affinis* to the period for narcosis and recovery except in some cases, where the same period was required for fixation in second to fifth stimulations.

Field Experiments on Mixed Population in AC Field

The fine meshed electrical seine net (Biswas, 1962) created AC field between the electrodes attached to the head rope and foot rope and the lines of electric field passed vertically through the entire water column (depends on the depth of water in the pond between 3 to 4 feet) between the electrodes, when the seine net moved from one side of the pond to the other.

Irrespective of the varieties, the catches were found to be more in electric seine than that of same net without electrical field. This may be due to the oscillotactic behavior of the fishes followed by hypnosis when they enter the vertical electric field (120 volt) produced by the electrodes attached to the head rope and foot rope of the seine net. Electrical resistance of the pond water (4200-18000 ohms/ml) played a major role in draining the input voltage (50-140 volts) between the electrodes. 24 to 48 per cent of excess catch (due to the effect of electric field) was observed in water resistance of 9550 ohms/ml (125 volt) and 9250 ohms/ml (120 volt).

Bottom dwelling fishes like, *Labeo calbasu, Channa* spp., *Glassogobius guiris, Mystus* spp, *Rhyncobdella aculeate, Clarias batrachus* and *Cirrhinus mrigala* were found to be more susceptible in electrical seine net than other varieties.

Fishes pf 20-25 mm, 15-20 mm, 10-15 mm and 5-10 mm sized fishes of the mixed population were caught in more numbers in electrical seine net than other size ranges.

Chapter 8
Comparative Effects of DC, AC and Pulsed DC on Fishes

Reaction of Fish in DC

On passing direct current in a fish tank, fishes will be found to swim towards the positive electrode, but before reaching it, they are usually stupefied and turn upside down. On switching off the current they recover and swim away.

Fishes exhibit several phases of reaction with rising current density. The first visible reaction of fish is that their bodies are vibrated by the current, being in parallel position to the current direction. In transverse position to the current, they turn their heads towards the anode.

Being in parallel position to the current direction, fish swim towards the positive electrode (*galvanotaxis*). The body-axis of the fish turns parallel to the direction of current, if fish is initially lying in a transverse position to the current direction. But the fish do not in all cases swim towards the positive electrode, unless a new pulse is given.

Reaction of Fish in AC

In alternating current (AC), the fish do not swim towards any one of the electrodes, as they do in the case of direct and interrupted DC, but they take a transverse position to the direction of current flow, between the two electrodes, in such a way that they tap off a minimum voltage (*oscillotaxis*).

The body color of the fish narcotized by alternating current fade owing to pigment contraction. Moreover, most of the fish (trout, carp, tench, catfish, etc), according to Scheminzky (1941) undergo a kind of "hypnosis" after the current is switched off.

They do not return immediately to normal swimming position, but stay for some times in a lateral or dorsal position. When hypnosis vanishes, the fish swim away as they do in the case of direct current.

Reaction of Fish in Interrupted Current

In interrupted current (direct current-pulses of rectangular or square wave form), or a pulsating current of triangular in form of a condenser-discharge or in the form of a half or quarter sinus wave, more or less heavy vibrations occur in the fish body, depending on the number of pulses per unit time. As soon as the pulse threshold for the electro-taxis is reached, the fish turns and swim towards the positive electrode. This happens with all the above mentioned forms of pulsating currents, except when the current is increased slowly and then decreased steeply. In the latter case, the fish turn towards the cathode, and narcosis occurs in the same was with DC.

Physiological Effect on Fish

Event though the visible external symptoms appear to be similar, when treating the fish with continuous DC or interrupted DC as well as AC, the investigated physiological effects are different. If the fish are treated with continuous direct current, a stimulant develops in the spinal cord, which according to Scheminzky and Kollensperger, causes the narcosis. When the current is switched off the stimulant disappears. Also, the galvano-narcosis effects diminish in the central nervous system. The galvano-narcosis can therefore, be considered as genuine primary paralysis, like chemical narcosis.

A polar stimulation of the nerves in the case of continuous and interrupted DC may be the cause of anodic reaction of fish in the electrical field, as revealed from the investigations of Halsband. According to Pfluger principle, the closing of the circuit has a stimulating effect on the nerves or muscles within the range of the cathode. The opening of the circuit, however, has a stimulating effect with in the range of anode. The reaction to the opening of the circuit is much smaller than the reaction to the closing of the circuit.

The behavior of the fish corresponds also to the Pfluger principle. When the circuit is closed, the fish is stimulated within the range of cathode and, consequently, it moves towards the anode. When the current is opened, the same effect occurs within the range of the anode and the fish moves towards the cathode, but only if the difference of the stimulative effect between the closing and opening of the circuit is great enough. This happens when the current rises slowly and declines steeply. A slowly increasing current has no stimulative effect. It is suggested that the nerve gets slowly used to the current when the current rises gradually (accommodation of the nerve). A steep decline, however, causes a stimulation great enough to induce the fish to swim towards the cathode. But if the current increases steeply and declines gradually, no movement towards the cathode occurs and the fish definitely reacts anodically. This proves clearly that the current pulses must have the proper shape of curve if an anodic effect on the fish is to be produced. The stimulus is felt by the ends of the sensitive nerves beneath the skin.

After the fish has received stimulation in the state of galvanotaxis, a reflex impulse occurs causing the fish to move towards the positive electrode. So the electric current is operational after the stimulation is received through the central nervous system. A removal of the spinal cord prevents the anodic movement. But it is to be distinguished between a coordinated movement towards the positive pole and a vibration of the muscle towards the anode caused by the current flowing in the fish. In the absence of spinal cord, a direct stimulation of the nerves and muscles by the electrical current, may however, may lead the animal with a jerk towards the anode, but no coordinated movement towards the pole can be observed. The anodic reaction is not caused by certain sense organs, as previously understood. The galvanotaxis can even be observed with decapitated animals and animals with cut lateral line.

The stupefying effect with AC and interrupted current can not be called a genuine paralysis. These types of current first cause a more or less heavy stimulation of the central nervous system, which leads mainly to titanic contractions of the muscles. This state of immobility is the consequence of cramp occurring in the muscles, which is caused by the superposition of numerous individual contractions.

The pulsating current has the greatest neuro-physiological effect. The alternating current takes second place, followed by the direct current. In spite of the great neuro-physiological effect of the pulsating current, Halsband was able to prove by investigations of the metabolism of the fish that pulsating currents have the least effect on the whole metabolism. He established this by measuring the basic oxygen consumption, breathing frequency and temperature of the intestine. Thus the after effect of the electrical flow of pulsating currents amounts on average to only 20 minutes, that of direct current to 60 minutes and that of alternating current to as much as 120 minutes. In other words, although the pulsating current has the greatest neuro-physiological effect, it has the smallest after-effect on the total organism,

The harmless effect of the electrical current on fish and their food animals has already been proved in several cases. Meyer Waarden, for instance, proved that even longer treatments with various types of currents influenced neither the viability, the growth, nor the increase in weight of carp, and Denzer and Riedel proved the harmlessness of the effect of direct current on the gonads of fish and their food animals.

Phenomena observed with fish are, as mentioned above, correlated with the natural symptoms from the general physiology of the nervous system. The nerve is stimulated by the electric current only when the circuit is closed or opened. A continuous current flow has no effect. However, there are differences between the vibrations caused by closing or opening the circuit. The threshold value for the closing vibration, that is, for the minimum voltage which effects a just visible reaction of the nerve, is smaller than that for the opening vibration.

Electrical Current Conduction through Fish in Water

The electrical effect on fish is possible in fresh water as well as sea water. The reactions are also the same, but the body voltage is some what greater in sea water than in fresh water. This is due to the fact that the electrical conductivity of the fish body is substantially smaller than that of the surrounding sea water. The fish-body

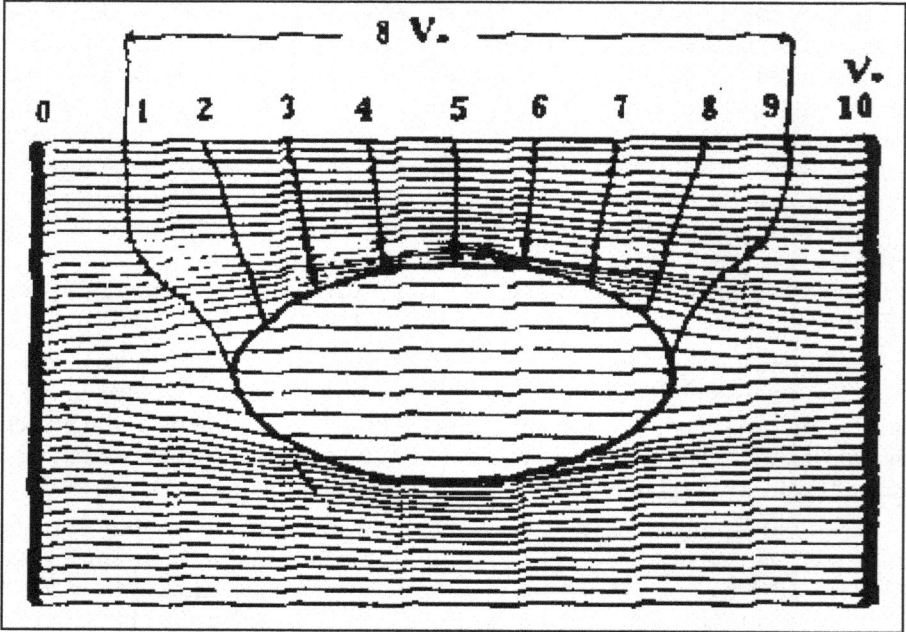

Figure 8.1a: The conductivity of sea water is better than the conductivity of fish body. The potential lines are distorted by the fish owing to its lesser conductivity. The body voltage increases to 8 volts as compared with 2 volts in fresh water and 5.3 volts in the "equivalent bath" (according to Holzer).

distorts the current field lines surrounding it, and the density of current in its vicinity, therefore is greater than other water (Figure 8.1a).

The conditions in an equivalent bath (conductivity of water is equal to the conductivity of the fish body) and when the conductivity of the water is smaller than that of the fish body, are shown in Figures 8.1b and 8.2.

The conductivity of sea water is on the average 500 times greater than that of fresh water. It is necessary, therefore, to use 500 times greater voltages in the case of direct current to produce a similar electrical field as in fresh water. But using pulsating current, the conditions in sea water are not unfavorable as compared with fresh water, because;

a) the actual effective value of the body voltage of larger marine fish is relatively small;

b) the smaller resistance of sea water permits short pulse-periods, when condenser discharges and suitable electrodes are used, which again means smaller average effects.

Types of Pulses

Due to higher conductivity of sea water, alternating and direct currents can not be used. The investigations revealed that it would be necessary io produce up to

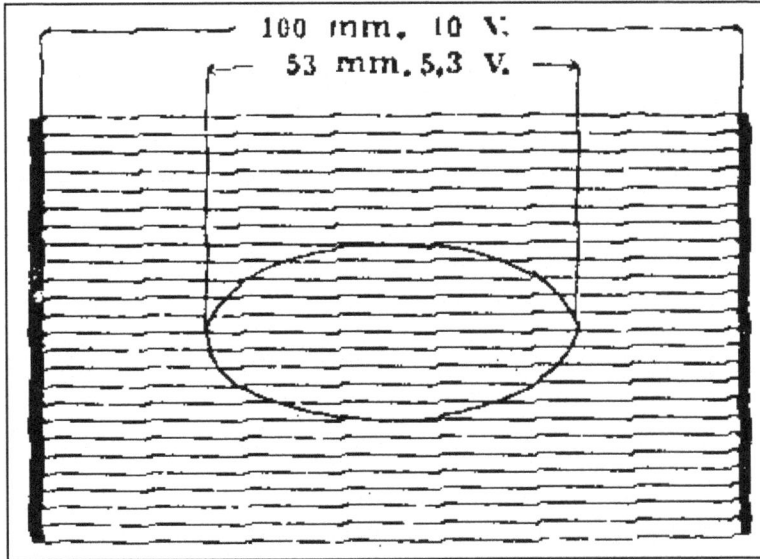

Figure 8.1b: Fish in an equivalent bath (fresh water with good conductivity), conductivity of water is equal to the conductivity of fish body. The fish taps off the potential lines corresponding to its length; 5.3 cm length of fish = 5.3 volts (according to Holzer).

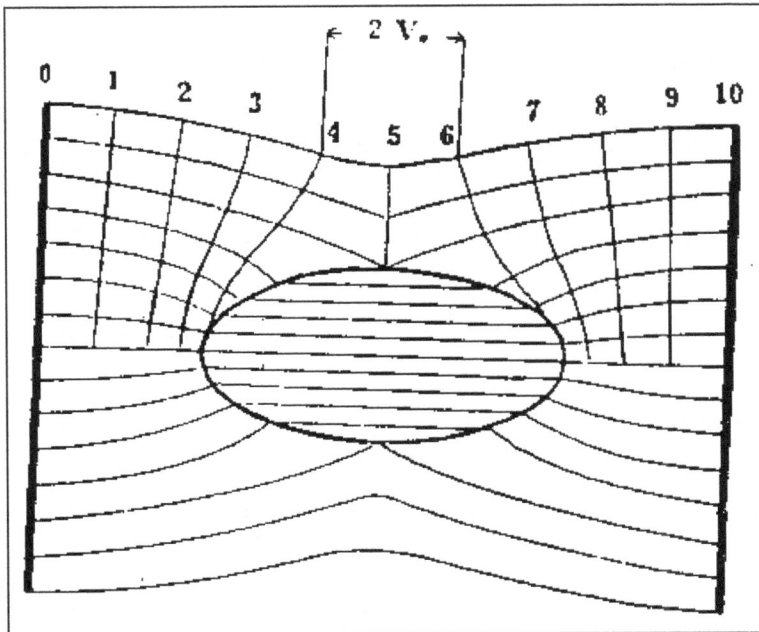

Figure 8.2: Conductivity of water is low than the conductivity of fish body. All potential lines are directed towards the body with better conductivity. The fish is thus satisfactorily influenced by electricity. The body voltage amounts to 2 volts as compared with the previous 5.3 volts (according to Holzer).

approximately 10000 kilowatts to obtain an electrical fishing range, of a radius of only 10 meters. In such a case it is only possible to operate with short pulses such as those of electric fishes. Only 200 to 300 kilowatts would then be required for the same effect. Leduc operated with a current of 100 periods, namely with impulses of one millisecond duration, followed by current interruptions of 9 milliseconds each. The body voltage of the fish in case of such currents requires an effective value substantially smaller than when normal direct current is used. The peak value of voltage used by Leduc amounts to 1.5 to 2.5 times the body voltage required in the case of normal direct current.

However, as these rectangular impulses with greater pulses of 10 to 20 kiloamperes (kA) with 1.5 to 3.5 kilovolts (kV) can not be produced economically, Kreutzer began, in the first experiment he made in electrical fishing in 1947, to use currents produced by condenser discharges with a steep rise and gradual decrease. It is to his credit that he recognized then that the shape of individual pulse is of great value.

The curve forms rectangular and unperiodic pulses are less suitable to electrical fishing than the quarter sinus curve and pulse from condenser discharge. The quarter sinus curve is specially used to block fish or to guide them in desired direction by means of movable gear. It is similar to the type of pulse produced by the electric fish. Curve form produced by condenser discharge (triangular form with steep rise and gradual fall) is the most effective form of pulse for attaining galvanotaxis. Most fish react to these pulses anodically.

Rate of Pulse

The number of pulses per unit time is decisive for the efficiency of the pulsating current, a fact already recognized by Kreutzer in his first experiments.The pulse rates at which fish are narcotized with the minimum of electrical energy, differ considerably for the various species as well as for fish of varying lengths. The values found for the narcotizing impulse thresholds are not to be considered as absolute, but as relative values, as they have been measured for certain sizes of fish and a certain water temperature. They change according to the temperature, the physiological condition and the size of t fish. The basic differences between the species of fish, however,are maintained.

Impulse Threshold

Narcotizing Pulse Thresholds for Freshwater Fishes (according to Halsband)

Species of Fish	Average Length in cm	Voltage of Individual Pulse for Narcosis in in Volts (1)	Narcotizing Pulse Threshold Values (Pulse rate/sec) (2)
Salmo gairdnerii	15-17	6.5	80
Cyprinus carpio	12-15	7.5	50
Amelurus nebulosus	14-16	11.5	40
Anguilla anguilla	20-22	13.5	50

Contd...

Contd...

Species of Fish	Average Length in cm	Voltage of Individual Pulse for Narcosis in in Volts (1)	Narcotizing Pulse Threshold Values (Pulse rate/sec) (2)
Idus melanotus	13-15	7.5	30
Acerina cernua	14-16	7.5	50
Blicca bjorkna	12-15	11.5	40
Phoxinus laevis	7-9	8.5	90
Gasterosteus aculeatus	6-7	13.5	100
Perca fluviatilis	12-14	9.5	70
Misgurnus fossilis	24-27	9.5	40
Tinca tinca	16-18	7.5	40

Size of experimental tank 35x22x22 cm; Size of electrode – 22x22 cm; Water temperature - + 15 degree Celsius

(1) Measured between the electrodes; (2) Triangular wave form rom condenser discharge

Narcotizing Impulse Threshold

Fish within a specific area remain stunned when they are affected by impulse rates which are above the threshold values required for narcotizing them. But if the impulse rates are kept below the limit which narcotizes fish, then the fish swim from a larger area towards the anode and, when they are closed to it, become narcotized.

With a low pulse rate the required electrical energy is reduced by half as compared with the purely stupefying method. This fact is of great importance to electrical fishing. And because different species of fish react differently to varying pulse rates, and as the threshold values of the pulse rates are known, it is possible to select the fish to be caught within a certain range by electricity according to size and kind.

The term "narcotizing pulse limit" was formulated by Kreutzer. He was also the first to recognize the importance of this idea for electrical fishing.

The importance of the size of the fish in relation to the narcotizing pulse limit can be explained as follows (Figure 8.3); each pulse causes a vibration of the muscles in the fish. If the next pulse occurs before the mechanical movement caused by the preceding pulse ends, the muscles are continuously stimulated and cramp sets in. This phenomenon is slower in developing in larger fish because larger masses of muscle have to be moved, and big fish do not, therefore, require such a quick sequence of pulses as small fish to produce cramp. Of course, a small fish can also be narcotized electrically at low pulse rates, but the length of the pulses must be substantially greater or the pulse voltage must be greater, and that would means deviating from the desired minimum of electrical energy, at which the necessary effect would just be reached.

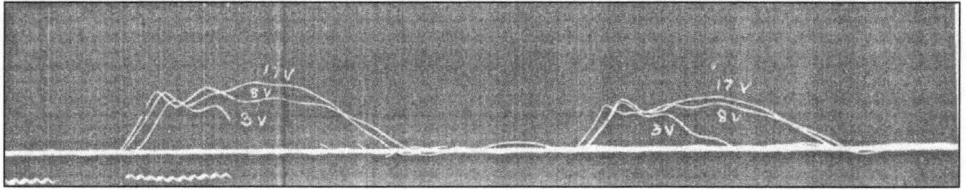

(a)

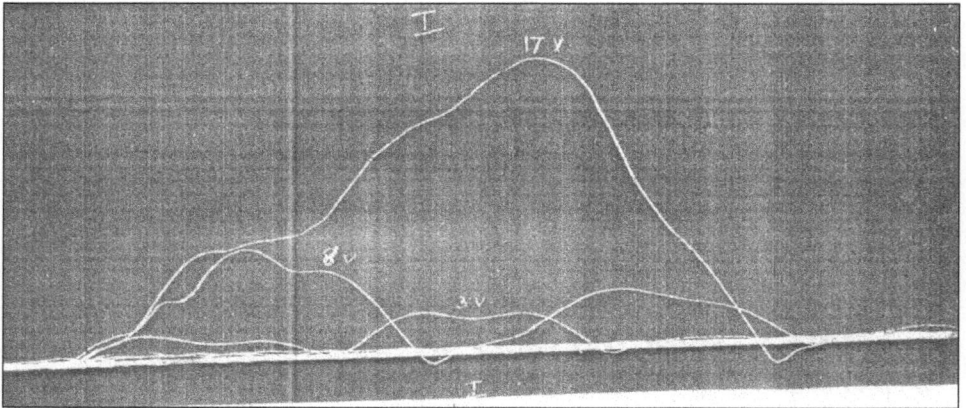

(b)

Figure 8.3: Effect of DC shock on the body muscle of (a) *Heteropneustis fossilis*; **(b)** *Tilapia mossambica.*

Electrical Stimulation and Duty Cycle

To produce a reaction in an isolated nerve, the current must have a specific minimum strength, generally known as rheobasis, and a specific minimum flowing period. The minimum flow is designated as an "effective period", which means that the electrical stimulation is produced only in that period. Any additional time used is wasted. The term "chronaxy" used in nerve physiology is the effective period of the double rheobasis. Nerve physiological experiments have shown that each strength of current or potential is related to a specific minimum flowing period of current so that a great number of effective periods have been found. If these values are classified in a system of coordinates, a curve develops, resembling a hyperbola. This curve indicates that the required potential falls continually with the increasing period of stimulation until a certain threshold value is reached. From this point the potential remains constant for all additional periods, even when infinitely long impulse currents are transmitted.

The curve runs in linear shape from a specific point onward. The turning point of the hyperbola indicates where the smallest consumption of energy takes place, and is the point where the most favorable conditions exist in relation to potential and flow of current for producing the electrical stimulation. This point corresponds to chronaxy. In nerve physiology, the chronaxy is thus, from the biological point of view, the stimulation effected with the greatest economy.

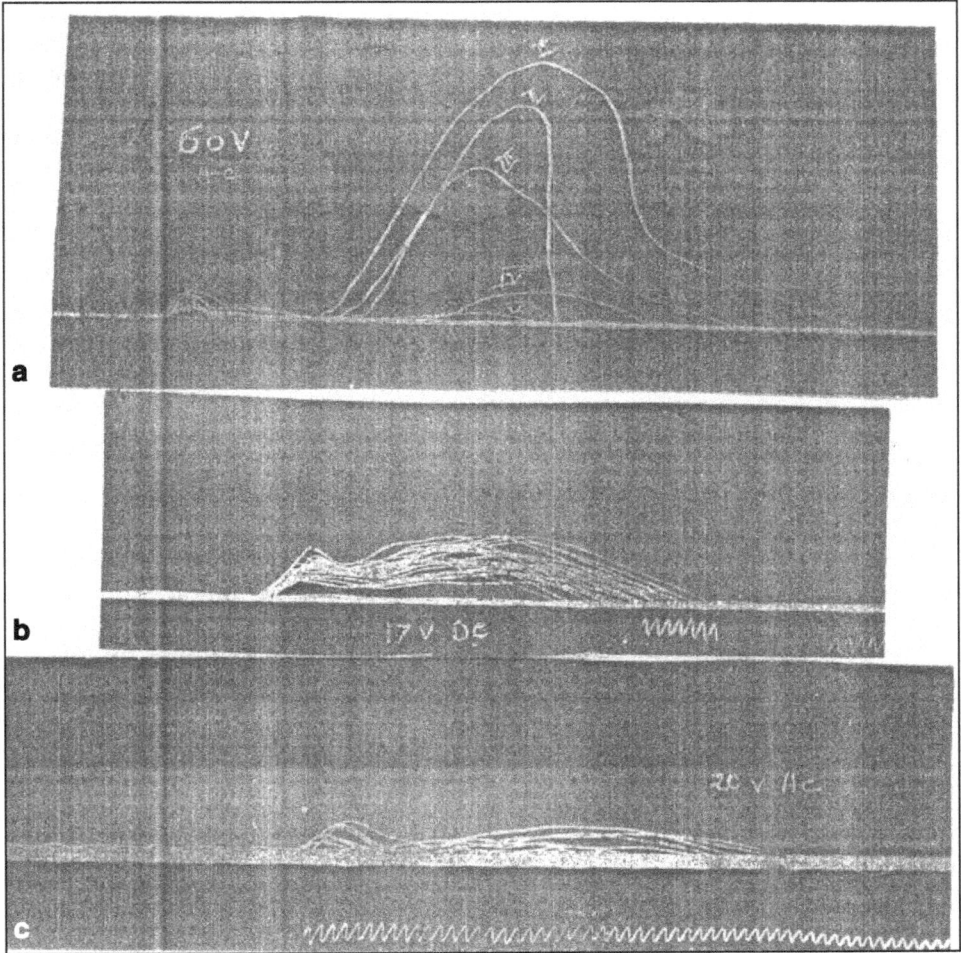

Figure 8.3a: Effect of repeated stimulation and fatigue on the body muscle of *Heteropneustis fossilis.*

a and c: AC shocks; b: DC shocks.

The principles demonstrated here for the isolated nerve are also found in the behavior of the fish as a whole, as proved by the investigations of Kreutzer and Halsband. Using pulsating currents, it was found that a continually greater potential is required for producing a stimulative effect, if the flow period of individual impulses is continually decreased. If the effective periods for all stimulative potentials are determined with the whole animal and expressed in a system of coordinates, a curve resembling basically that known from nerve physiology is obtained. The figures of the potentials and values of the pulse lengths obtained for the whole fish do not quite correspond to the values found for the isolated nerve. On determining the required potential values from the several points of the curve, it will be recognized that the consumption of energy is large with long impulse periods, and that it also increases with short pulse periods. Between these two extremes exists a point of smallest consumption of energy, the minimum of effect, where the most favorable conditions concerning the potential and the impulse period exist for causing a certain reaction. These points of smallest consumption of energy, showing but little variation for the different species of fish, are of great importance in designing impulse fishing gear for electrical fishing, in order to obtain economically useful values especially at sea.

From the technical point of view, the current flow of a single impulse is the flowing period of a rectangular impulse (Leduc current). The average current flow or the so-called half value time is chosen as the time constant for condenser discharges It results from the flow period of the corresponding rectangular impulse X 0.37.

As stated by Kreutzer previously, the half value time may not be below one millisecond when condenser dischsrge is used. But as he had already mentioned, this flow period of the current may possibly be further diminished. Actually, these flow periods are considerably less. The optimum pulse rates required for taking the fish fluctuate within a species; they also depend on the size and physiological condition of the fish. Also there exists a relation between the impulse period and the impulse rate. Bary (1956), for instance, observed with sea basses that, if the impulse period is extended beyond the optimum, lesser energy and smaller impulse rates are required to cause certain reaction in the fish. When using an optimum impulse rate, Halsband found for trouts of medium size an optimum impulse period of 0.6 millisecond (half value time = 0.25 millisecond) for reaching the stage of galvanotaxis. Bary mentioned an impulse period of 0.5 millisecond for reaching that stage in sea basses of medium size.

Narcotizing Pulse Thresholds for Sea Fish

Species of Fish	Average Length in cm	Voltage of Individual Pulse Required for Narcosis	Narcotizing Pulse Threshold Values (Pulse rate/sec)
A) According to Halsband			
Myxocephalus scorpius	15-19	6.2	40
Zoarces viviparous	17-21	10.0	60
Pleuronectes platessa	23-26	10.0	30
Ciliata mustela	18-17	11.5	60
Osmerus eperlanus	18-22	12.0	50
Platichthys flesus	16-20	11.5	40
B) According to Kreutzer			
Herring of medium size			45
Cod of medium size		25	
Red Tuna (200-300 kg)			7-10

Size of experimental tank : 35x22x22 cm; Size of electrode : 22x22 cm.

Water temperature - + 15 degree Celsius.

Chapter 9

Electric Stimulation on Heart Beat and Fish Body Muscle

Electric shock response in man (Dalziel and Lee, 1968) has shown that electric current is the best measure of the strength of the sensible shock effect. Similar results were observed in fishes also (Cuinat, 1967). The primary function of the electro-fishing system is to establish an electric current in water near the fish. Depending on the ratio of fish and water conductivity, a portion of this current will pass through the fish and elicit the desired response. The skin of fish with numerous sensory organs is an important receptor for physical and chemical stimuli (Brett, 1957). Biswas and Karmarkar (1979) attempted to measure the electrical resistance of the skin and body of fish, the contraction of fish body by direct current and also the effect of electrical stimulation on the heart beat in fishes.

The electrical resistance of the skin and fish body at different points were measured by using a dermometer described by Whelan (1950), which had a range of 1000 to 45000000 ohms at about 5 per cent accuracy. Live fishes were measured for length and maximum width and kept in between the folds of filter paper to remove water from the surface. One end of the silver electrode was connected to the instrument and the other end pressed against the fish, the instrument switched and adjusted to 2 μA. The resistance of fish was directly read from the voltmeter and multiplied by 10 or 100 depending on the range selected. The resistance between different body parts, namely, mouth to top of head, mouth to mid-dorsal fin, mouth to caudal peduncle and mouth to vent were measured with platinum electrodes.

To record the contraction of the body in *Heteropneustis fossilis*, its peduncle was clamped firmly, upper jaw connected through a thread to a lever provided with a writing pen. The writing pen touched the black surface of a smoked paper wound round a cylindrical drum moving at constant speed (kymograph). The pen drew a horizontal line (abscissa) on the smoked paper when the drum rotated. The brain together with a portion of a spinal cord was destroyed by inserting and rotating a needle. The lever turned on the fulcrum, jerked up when the fish body received electrical stimulation and the pen drew a curve on the smoked paper. The time was furnished by a tuning fork vibrating at 100 times per second.

The effect of electrical shock on the heart rate by the applied underwater field was recorded graphically on a constant slow moving drum. The fish after destroying the brain and spinal cord was firmly clamped upside down at the bottom of the experimental tank, where a homogeneous electrical field was created by two copper plates placed vertically along the tank extremities. The heart of the animal was then exposed and the ventricle was connected to a lever, fitted with a writing point, which in turn touched the blackened surface of a tracing paper wound round a drum, moving at a slow speed. After recording normal heart beat for some time, the animal as a whole was subjected to electric shocks of different intensities and the heart rhythm was recorded for some time. The recordings continued even after switching off the current till the heart return to normal rhythm. Currents of different forms and stimulations of different types were tested on *H. fossilis* and *Tilapia mossambica*. Visual observations of the effect of electrical shock on the heart rate of *Macrobrachium rossenbergii* were also recorded.

The spinal cord of *Puntius ticto, H. fossilis* and *T. mossambica* were severed first behind the supraoccipital and were allowed to rest for two hours before exposing them to electric shocks. The set up and the procedure were the same as described earlier. Observations were made for the sensitivity and response in these animals and the minimum field strength required to elicit response in them.

Body Resistance and Muscular Contraction

The electrical resistance of *Mystus aor* was the lowest (6x1000 – 7x1000 ohms) and of *M. rosenbergii* the highest (24x1000-25x1000 ohms). The result of tolerance test showed 100 per cent mortality in *M. aor* and *Cirrhinus mrigala* within 25 to 35 and 65 to 75 minutes respectively after exposure to current. Sixty per cent of *Notopterus notopterus* and 40 per cent of *M. rosenbergii* died after 85 to 105 and 92 to 120 minutes respectively. Twenty five per cent of *Labeo rohita and Catla catla* were found dead within 115 to 135 and 218 to 255 minutes respectively after exposure to current. An increase in the latent period of 20 milliseconds under 3 volts was recorded over that of 8 and 17 volts in *H. fossilis*. A sharp rise (contraction) with two peaks and a gradual fall (relaxation) were observed at all the shock intensities tested. The contraction phase lasted for 80, 160 and 180 milliseconds for 3, 8 and 17 volts respectively. The corresponding relaxation phases were 120, 140 and 110 milliseconds. In *Tilapia mossambica* two distinct curves with double peaks and a latent period of 90 milliseconds were recorded for DC of 3, 8 and 17 volts. At 3 volts the amplitude of the second curve was more than the first one. But the amplitude of the first curve increased with the intensity of current (8 and 17 volts). The contraction phase lasted for a higher period in 7 volts than the relaxation period.

At 20 volts the curve of muscle twitch in *H. fossilis* showed a sharp rise and fall with two small peaks at the top. When raised to 40 volts, two peaks along with a wide valley at the top was noticed. At 60 volts, a single peak of highest amplitude was drawn. The contraction phases remained for 190, 450 and 425 milliseconds for 20, 40 and 60 volts respectively. The corresponding relaxation phases, however, lasted for 20, 30 and 185 milliseconds respectively. The contraction of body muscle of *T. mossambica* due to alternating current of 20, 40, 60 and 80 volts revealed the formation of two curves during each stimulus. The height of the second curve was more than the first and a double peak was noticed in the second curve at 80 volts.

The latent period was prolonged by 15 to 50 milliseconds from the third stimulation onward. The length of contractions were also increased irrespective of DC and AC. After an initial rise during the first few contractions, the height of the curve reduced uniformly. The relaxation phase was prolonged with the number of stimuli and after 30 to 40 contractions, the writing lever took several milliseconds to return to the baseline.

Effect of Electrical Stimulation on the Heart Rate

The heart rate of *M. rosenbergii* when exposed to field intensity of 0,0792 to 0.2376 µA/sq mm for 15 to 180 seconds, showed a decrease of 10 to 12 per minute during the shock. The heart beat decreased further under current intensities of 0.792 to 1.98 µA/sq mm till the heart stopped during narcosis. Increasing shock intensity suppressed heart beat (61-116 per minute), but accelerated the rate to 146-212 per minute immediately after recovery from narcotic condition in some cases.

T. mossambica showed slowing of heart beat soon after exposition at DC 0.8 µA/sq mm and 1.2 µA/sq mm caused momentary cardiac slowing, but afterwards came back to normal rhythm prior to shock treatment.

H. fossilis when subjected to increasing intensity of alternating current, its heart curve showed initial rise during first few contractions followed by uniform decrease till the current intensity reached 0.55 µA/sq mm. At a field intensity of 0.75 µA/sq mm irregular fluttering of heart was recprded, which regained normal rhythm when the current was switched off. Gradual cardiac slowing was observed when *H. fossilis* exposed to sharp current rise of 0.35 and 0.55 µA/sq mm and heart stopped in the relaxing phase after 35 and 11 contractions respectively. Exposure at higher current intensity (0.75 µA/sq mm) caused stoppage of heart beat after three contractions.

Stimulating the heart with current intensities at 0.7 µA/sq mm and 1.1 µA/sq mm in a media of calcium chloride (1:1500) having a resistance of 8x1000 ohms/sq cm, the relaxation after each contraction became more and more incomplete, until finally the heart stopped in a tonically contracted condition. In potassium chloride solution of identical concentration and resistance, the tonic condition of the heart increased with 0.2 to 0.75 µA/sq mm current and the contractions lasted as long as the current flow continued. During treatment with current intensities of 0.2 µA/sq mm and 0.35 µA/sq mm, however, the amplitude of heart beats were more when compared to shocks of higher intensities.

In order to control fish by electricity in freshwater, a high field strength is required, because the conductivity of the fish is higher than that of water (Bary, 1956). When a fish comes to an electric field, the charge induced on the fish is proportional to the cube of its body length (Kuroki, 1952). According to Holzer (1931), when the conductivity of fish body is higher than the surrounding water (fresh water) all the potential lines are directed towards the body with better conductivity and the fish is thus satisfactorily influenced by electricity. The knowledge of electrical resistance of the fish body is thus important in initiating the responses and determination of reaction thresholds.

Halsband (1968) measured the electrical resistance of trout body through water media. Biswas (1979) determined the electrical resistance of fish body by the dermometer and represented as ohms/sq cm of body and between different points of body as studied by Denzer (1956) for rainbow trout. The electrical resistance (ohms/sq cm) was found to be lowest in case of *M. aor* (devoid of scale) and highest in *M. rosenbergii* (with a chitinous shell). As observed by Denzer (1956) the resistance between mouth to caudal peduncle of all the test fishes was highest than between any other parts of the body which is due to the largest body extremities of the animal. A fixed amount of current (25 μA) was allowed to pass through their body by direct contact for a period of 10 seconds. The contact voltage varied from 1.4 to 2.4 volts depending on the species. The severity of shocks (as observed by the percentage of death and the time required for death) indicate that the test fishes are sensitive to DC and were affected by electric shocks in the order of *M. aor, C. mrigala, N. notopterus, L. rohita* and *C. catla*.

Effect of Electrical Stimulation on Skeletal Muscle

The skeletal muscle, or its nerve when stimulated by a single induction current resulted in a single short sharp contraction followed immediately by a relaxation (Starling, 1947). To determine the time relations of muscle contraction it was necessary to employ the graphic method. The mechanical changes that the fish as a whole had undergone by DC and AC were represented by kymograph tracings along with a time record.

Starling (1947) demonstrated that the simple muscle twitch in frog consists of three main phases, namely, the latent period(during which no apparent change takes place in the muscle), contraction (a phase of shortening) and a phase of relaxation. When the whole body of *H. fossilis* and *T. mossambica* were stimulated by DC of varying intensities, the latent period was delayed in currents of low intensities (Biswas, 1979) According tp Meyer-Waarden (1957) the effective period inducing narcosis (relaxation of muscle) in fishes was inversely proportional to the intensity of the field. The occurrence of second contraction before attaining complete relaxation in both the test fishes can be explained by Pfluger's principle (Pfluger, 1853), that is, the closing of the circuit has a stimulating effect on the nerves or muscles within the range of the cathode and the opening of the circuit has also a stimulating effect within the range of the anode and the reaction to the opening of the circuit is much smaller than the reaction to the closing of the circuit. But in case of DC on *T. mossambica*, the amplitude of the second curve (due to opening of the circuit) was found to be greater in 3 volts when compared to currents of 8 and 17 volts.

In AC also, double curves for each stimulation was obtained in case of *T. mossambica*. But in *H. fossilis*, two peaks at 20 and 40 volts and a single peak of highest amplitude during 60 volts could be explained by summation effect of two stimuli during closing and opening of the circuit. When the intervals between two currents were shortest, a greater response was observed with a single peak. But during longer intervals between the first and second shocks, as in case of 20 and 40 volts, the excitatory condition of muscle was maintained, so that instead of forming two isolated muscle twitch, a second contraction occurred before the fading of the effect of previous stimulus forming two peaks in one curve.

The effect of repeated stimulation by both DC and AC on *H. fossilis* showed that the latent period was prolonged, with the length of contractions. The AC and interrupted DC caused heavy stimulation of the central nervous system which led to titanic condition of the muscles (Meyer Waaeden, 1957). According to him, this state of immobility is the consequence of cramp occurring in the muscles caused by the superposition of numerous individual contraction as observed with *H. fossilis* with AC.

Effect on the Heart Beat

William (1885) demonstrated in the eel that almost any sort of peripheral stimulation caused inhibition of the heart beat and this appeared to be generally true in unanesthetized fish. Cardiac slowing in *M. rosenbergii* was observed during the application of DC depending on the current intensity. A slight hyper-activity of the heart was noticed afer recovery from narcosis. Momentary cardiac inhibition of *T. mossambica* was also observed in DC of varying intensities (0.12 to 1.2 µA/sq mm). Studies in man and animals with AC indicated the occurrence of ventricular fibrillation (Novotny and Priegel, 1974). The heart of *H. fossilis* when exposed to AC with slowly rising intensity, exhibited staircase phenomenon (Starling, 1947) at current densities of 0.1 µA/sq mm, cardiac slowing at 0.55 µA/sq mm and irregular fluttering at 0.75 µA/sq mm. In sharp current rise, cardiac arrest occurred in relaxation phase even at 0.35 µA/sq mm. Switching off current flow immediately after stoppage of heart beat, the heart regained the normal rhythm once again indicating that the period of exposure in relation to current intensity is responsible for inihibition of heart rate (Figure 9.1).

The inhibition effect of calcium chloride solution and the accelerating effect of potassium chloride on heart muscle (Starling, 1947) was confirmed from the tests also, where the heart of *H. fossilis* stopped in a tonically contracted condition, when stimulated through calcium chloride solution and continued to contract in the tonically contracted state even during the current exposure while stimulating through potassium chloride solution (Figure 9.1).

In the nervous control of locomotion and of movements, the spilal cord clearly plays an important role. Pfluger (1853) severed the spinal cord of eel (*Anguilla anguilla*) immediately behind the medulla oblongata and described various reflexes shown by

Sharp Rise

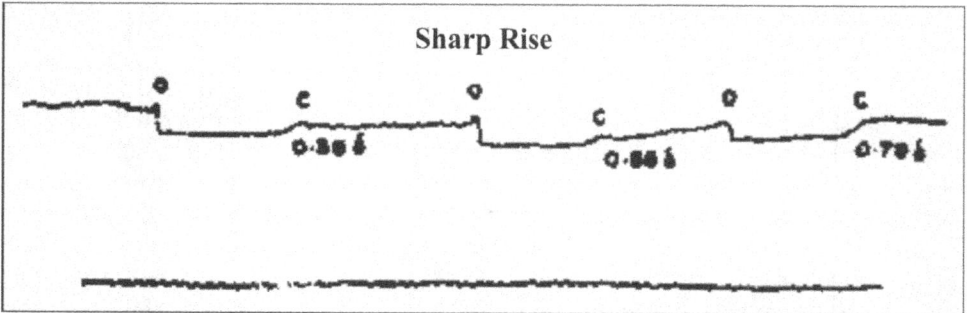

Figure 9.1: Effect of sharp increase in AC current on heart beat of *H. fossilis*.

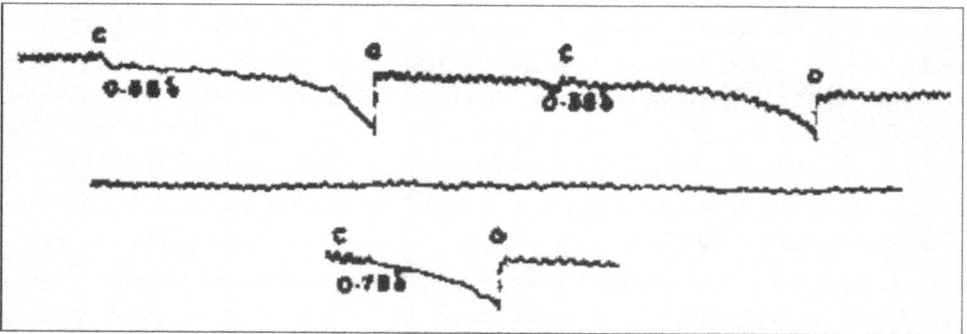

Figure 9.2: Effect of AC current on heart beat of *H. fossilis* (Medium- potassium chloride 1:1500).

the posterior portion. Von Holst (1934) concluded that after the spinal section only the anterior part of teleosts (*Carassius carassius* and *Cyprinus carpio*) showed rhythmical swimming movements, the hind part being motionless excepting when stimulated.

The fishes (*P. ticto, H. fossilis* and *T. mossambica*) after spinalization behaved in the similar way as described by Von Holst (1934) when exposed to DC shocks of varying intensities. The occurrence of anodic taxis, anodic narcosis and anodic curvature in *P. ticto* and *H. fossilis* were observed in a much lower thresholds as that of intact animal (Biswas, 1974). The un-coordinated movement of *T mossambica* towards the negative electrode in lower threshold than the intact animal (Biswas, 1974) confirmed the findings of Cuinat (1967) that the sharp spinal section allowed the persistence of galvanotaxis at the threshold of the intact animal which became unbalanced through the functional suppression of the cerebellum (spinalization).

Chapter 10

Factors Affecting the Reaction of Fish in Electric Field

Field Strength and Body Voltage

As stated in earlier chapters, reactions of fishes in successive stages occurred with the rise of current density depending on its position in electric field. The first reaction, galvanotaxis and galvanonarcosis in fishes occur only when the current density in the surrounding field has reached a specific value depending on length of fishes, conductivity of water, nature of current and varieties of fishes. The three reactions occurred on reaching the threshold potential drop between the head and tail (body voltage) besides the attainment of optimum current density. The threshold values of both current densities and body voltages are interrelated and fishes exhibited these reactions in successive stages due to the cumulative effect of density of current in the field and difference of potential created between the head and tail of the fish which varied inversely with the length of the fish. Not only the absolute current densities for these reactions reduced with the increase in fish length, but the ratio between these reactions also decreased with the increase in fish length.

Threshold current density and body voltage for different reactions of *Salmo irideus* in relation to length; Impulse frequency 82/sec; Impulse duration – 5 ms; Pause duration-7.2 ms (According to Biswas, 1971).

Length of Fish in mm	Mean Body Voltage in Volt			Mean Current Density in µA/sq mm		
	I	*II*	*III*	*I*	*II*	*III*
80-84	0.840-0.984	2.40-3.16	4.1-5.1	0.171-0.208	0.70-0.92	1.1-1.3
85-90	0.585-0.722	2.34-2.72	3.8-4.7	0.063-0.132	0.60-0.76	1.1-1.5
135-140	1.470-1.957	3.01-3.38	3.5-3.8	0.180-0.284	0.47-0.57	0.6
155-160	1.317-1.580	3.27-5.60	5.2-5.9	0.103-0.161	0.47-0.67	0.8-0.9
162-165	1.155-1.312	2.97-4.46	5.3-5.9	0.077-0.115	0.37-0.47	0.8-0.9
166-170	1.008-1.360	2.12-4.25	4.0-6.8	0.057-0.098	0.23-0.57	0.5-0.9
175	1.4	2.98	5.6	0.098	0.35	0.7

The threshold current densities required to initiate three reactions were found to be different irrespective of different lengths and water conductivity. Among *Salmo irideus, Idus melanotus* and *Cyprinus carpio*. The threshold values for *S. irideus* were minimum and those for *C. carpio* the maximum which was in conformity with the observation of Meyer Waarden (1957), who found that these values were specific for species and size of fish.

The reaction thresholds (for different stages of reactions) for *Puntius ticto* of 61 to 100 mm, *Heteropneustis fossilis* of 106 to170 mm and *Tilapia mossambica* of 91 to 200 mm revealed that the mean threshold current densities required for the first reaction were 0.04 to 0.08 µA/sq mm for *P. ticto*, 0.06 to 0.15 µA/sq mm for *H. fossilis* and 0.15 to 0.45 µA/sq mm for *T. mossambica*. The corresponding values of threshold current densities for galvanotaxis of the above mentioned species were 0.12 to 0.182 µA/sq mm, 0.12 to 0.28 µA/sq mm and 0.44 to 0.9 µA/sq mm respectively. They entered into a stage of galvanonarcosis in current densities 0.16 to 0.3 µA/sq mm, 0.14 to 0.3 µA/sq mm and 0.61 to 1.08 µA/sq mm respectively. The off balance swimming in the fourth stage occurred in some of the fishes in the threshold range of 0.16 to 0.22 µA/sq mm in the case of *P.ticto*, 0.14 to 0.18 µA/sq mm in the case of *H. fossilis* and 0.61 to 1.0 µA/sq mm in the case of *T. mossambica*. The tetanus and the muscular rigidity had been set up in them in current densities of 0.2 to 0.35 µA/sq mm, 0.18 to 0.33 µA/sq mm and 1.12 to 2.18 µA/sq mm respectively (Biswas, 1974).

A correlation existed between the potential gradient and the density of the current in an uniform homogeneous field, and the relation between the two parameters were fairly constant for every set of tests and depended on the conductivity of water media. A potential gradient of 10 to 17.8 millivolt per cm. could effect the field intensity of 0.01 µA/sq mm.

Lamarque (1963) concluded that the fish behavior in an electric field depends essentially on the potential gradient or even better, potential difference between head and tail. Chmielewski (1965) put much stress on the determination of head to tail voltages necessary to produce the first reaction and electronarcosis, since an electric field affects fish more strongly when head to tail voltage is higher. Since the head to tail potential difference has a direct correlation with the voltage gradient of the field, the potential gradients for bringing out the response were determined in case of *P.*

ticto, H. fossilis and *T. mossambica*. Dijkgraaf *et al* (1963 and 1968) reported the response of *Scyliorhinus canicula, Raja clavata* and *Ictalurus nebulosus* in potential gradients of 0.01 microvolt/cm in the case of first two species and 30 microvolt/cm for the third.

Lamarque (loc. Cit.) has stated that the threshold potential gradient for initial reaction to final stage of shock varied between 100 to 1250 mV./cm in case of eel, trout, skate, plaice, sole, golden fish, carp, tench etc.

The threshold voltage gradient for the three species for different reactions indicated that galvanotaxis and galvanonarcosis of *P. ticto* averages 2.7 and 3.9 times that for the first reaction. The corresponding values were 1.9 and 2.7 times for *H. fossilis* and 2.5 and 3.1 times for *T. mossambica* respectively (Biswas, 1974). A similar observation has been reported by Lukashov *et. al* (1963), where he has observed that gradually established body tension inducing narcosis averages 1.4 times that for electrotaxis.

An abruptly established potential difference induced *P. ticto, H. fossilis* and *T. mossambica* to enter into clearly demarcated reactions at every stage. An unmistakable anodic taxis followed by true narcosis have been observed in fishes during closure of the circuit with a lesser percentage of anodic curvature across the field and off balance swimming followed by tetanus (Biswas, 1974).

Halsband *et. al* (1960) stated that a steady direct current had no effect on fish at low potential values. Stimulation of nerves occur only in response to a change in circuit. The rate of decrease in threshold current densities in quickly made up potential over slowly increasing ones were 76.19 per cent for taxis, 90.01 per cent for narcosis and 45.18 per cent for tetanus in *P. ticto*. For *H. fosss* the rate of decrease in reaction thresholds were 34.61 per cent, 10.99 per cent and 26.99 per cent respectively for first reaction, galvanotaxis and galvanonarcosis. The corresponding values for *T. mossambica* were 53.87 per cent, 79.7 per cent and 73.25 per cent respectively. The increase of 43.4 per cent reaction threshold during the first reaction in the case of *P. ticto* was due to changing its position parallel to isopotential lines with switching on the current at sub-threshold values for taxis, where the fish tapped off a minimum voltage drop between the head and tail (Biswas, 1974).

The above findings goes to support Halsband's statement that current with a slow rise has no stimulating effect. The nerves become accustomed to the current when it rises slowly (nervous accommodation) and consequently a higher threshold values of potential gradient and current densities will be required to induce reactions even in a lower percentage of fishes for specific movement.

Fish Species

Different species of fish also react differently to varying pulse rates. The narcotizing pulse limit of *S. irideus, C. carpio, G. aculeatus, T. tinca* and *S. fario* were 82, 78, 65, 78 and 73 respectively in impulse DC of square wave form irrespective of conductivity of water. With impulse DC of rectangular wave form best anodic effects during galvanotaxis and galvanonarcosis were observed in frequencies of 34, 25 and 25 in case of *S. irideus, c. carpio* and *I melanotus* both in lower and higher conductivities of water.

The pulse threshold values needed to induce the reactions (sensing of current, galvanotaxis and galvanonarcosis). The threshold values of the pulses are constant and reproducible for fish of the same species and size. No reaction can be observed from pulses below that given for the first reaction.

Values (µA/sq mm) of the current density for reactions of freshwater and sea fishes (Halsband)

Species of Fish	Length cm	Weight in g	Values of µA/sq mm for Reactions		
			I	II	III
Gasterosteus aculeatus	5-6	3-4	0.48	1.88	3.94
Nemachilus barbatulus	10-12	7-9	0.08	1.55	2.43
Pleuronectes platessa	20-22	95-100	39.91	44.34	70.95
Agonus catophractus	8-10	14-16	18.18	50.99	115.00

The response of fish to under-water electric field are known as functions of field strength (Morgan, 1951; Groody *et. al*, 1952, Harris, 1953; Matsuche, 1955 and Bary, 1956). Their reactions were, therefore, observed by changing the field strength by increasing the applied voltage to the electrodes. Since fishes were not attracted by AC field (Meyer Waarden, 1957 and Levin, 1957), tests were carried out with DC and pulsed DC (Biswas, 1977).

A number of investigators have described the effects of DC and pulsed DC on fishes and prawns. The identification of the various reactions and the mechanism of reaction of fish in an electric field have been explained by authors like, Scheminzky and Kolensperger (1938). Holzer (1932), Cattley (1956) and Denzer (1956). However, a diversity of opinion between them on the behavior reactions, and the mechanism of reaction existed due to;

a) multiplicity of studies carried out under different experimental conditions inaccurately described;

b) various methods of stimulation and

c) terminology for reactions which often give rise to confusion.

The matter has been reviewed by Nusenbaum and Faleeva (1961), Halsband and Meyer-Waarden (1960) and Lamarque (1963), where suitable terminology of various current characteristics, electric field characteristics and behavior and physiology of fish in an electrical field have been standardized and defined by them.

It was established that the reaction of test animals (Nine varieties of freshwater Indian fishes and one variety of freshwater prawn) appeared in several phases, with a gradually raising current density in both DC and pulsed DC and these results are in agreement with the observations of all the earlier investigators (Morgan, 1951; Dickson, 1954 and Halsband, 1960). In DC field, perception of the surrounding electrical field evidenced by the stretching of fins, vibration of the body with occasional jerks (defined by Morgan, 1951 as fright reflex) were noticed in all the fishes and prawns in different degrees.

The second reaction, characterized as anodic taxis, when the fish swam towards the positive electrode (Halsband, 1955) was prevalent in the case of *Labeo rohita, Catla catla, Notopterus notopterus, Channa punctatus* and *Wallago attu*, when they were in parallel position to current direction. The movement of *Mystus aor* towards the negative electrode in this stage, was found to be in agreement with the observations of Denzer (1968), Lamarque (1963) and Biswas (1974), when similar responses were noticed in *Tilapia nilotica,* skate, plaice, sole and *Tilapia mossambica.*

The autoadjustment of *Macrobrachium rosenbergii* by curving their body and by *Labeo rohita, Cirrhinus mrigala, Channa punctatus, Wallago attu, Mystus aor* and *Carassius auratus* by orienting their body axis perpendicular to current lines to tap off minimum potential difference between their head and tail was also described by Kreutzer (1949) and Cattley (1955).

Immobilization of all the fishes and prawn during the third reaction, defined as galvanonarcosis by Halsband (1959) was considered as genuine primary paralysis, similar to that of chemical narcosis (Meyer-Waarden, 1957), where the slackening of body miscles along with cessation of respiration was noticed in the subject animal. According to Scheminzky and Kollensper ger (1938). A stimulant developed in the spinal cord, when the fish was treated with continuous DC caused narcosis. When the current was switched off the stimulant disappeared.

As soon as the organism rolled over on one side and was incapable of movement, the flow of current was stopped and the animal was allowed to revive back to its normal condition, The death of all the specimens of *Mystus aor* after the stoppage of current flow indicated their highly sensitivity to DC shocks over all other varieties of fish and prawn (Biswas, 1977).

Fish Size

It is significant that these pulse-threshold values decrease with the increasing size of the fish, although the proportion among the pulse-threshold values remains approximately constant.

Relation of body length to pulse threshold values for reactions in *Phoxinus laevis* (According to Scheminzky).

Length Group in mm	Median Length in mm	Current Density Thresholds (µA/sq mm) for		
		I	*II*	*III*
15-20	19	0.19	3.30	4.74
21-30	24	0.18	3.13	4.74
31-40	36	0.13	3.32	3.94
51-60	57	0.11	1.43	2.50
61-75	66	0.13	1.28	2.15

Holzer (1932) calls the voltage tension between head and tail, which is required to produce one of these reactions, "body-voltage". It amounts, in the case of trout for instance, to 0.4 volt for the first reaction, to 1.2 volts for the second reaction, and to 2 volts for the third reaction. Since large fish receive a greater voltage in the water than smaller ones, they can be influenced more quickly and by relatively smaller voltages. Fish which receive a lower voltage than that required for electrotaxis can escape from the field.

The requirement of reaction thresholds were found to be inversely proportional with the increase in fish length in all the three species (*P. ticto, H. fossilis* and *T. mossambica*). Several reasons can be put forward, but the most evident one is that large fish have longer nervous structures. Short nerves in an electric field are excited at a higher value of current than long nerves. (Laugier, 1921). Scheminzky *et. al* (1938) stated that not only the absolute current densities decreased with increase in size of fish, but also the ratio of current densities for the reactions decreased with increase of fish size. *Phoxinus laevis* of 19 mm median length at a mean ratio of current densities of 1:17:25 responded for the three reactions, while fishes of 66 mm size exhibited those reactions at a ratio of current densities of 1:10:17. But Indian fishes, like, *P. ticto, H. fossilis* and *T. mossambica* did not follow the similar pattern. An increase in ratio between their reaction thresholds with the increase in fish length were invariably noticed even though the absolute current densities for the reaction were lowered with the increase in fish size.

The length of fish was one of the most important factors which influence the changes in the rate of threshold current intensities and occurrence of different reactions. The reaction thresholds for *P. ticto* tend to indicate a negative correlation between the threshold current densities and length of fish in direct current of gradual and sharp rise. The correlation coefficient ® were -0.96, -0.99, -0.95 and -0.97 for the first reaction, galvanotaxis, galvanonarcosis and tetanus with gradual current rise. The corresponding values with steep current rise were -0.99, -0.307, -0.308 and -0.97 respectively.

The value of "r" in case of *H. fossilis* for the former three reactions as stated above were -0.66, -0.98 and -0.93 for slowly rising current and -0.98, -0.97 and-0.97 respectively for sharp current rise.

In case of *T mossambica* the values of "r" for the above three reactions were -0.99, -0.98 and -0.99 in slowly increasing potential. These values in abruptly established potential however, were -0.97, -0.98 and -0.99 respectively (Biswas, 1974).

Groody *et.al* (1952) observed that the current density required to produce directional swimming in Pacific sardines varied inversely with the size of the fish. Norman and Laukashkin (1954) noticed that optimal average current density required for galvanonarcosis and control of movements of *Sardinops caerulea* increased with decrease in length of fish. This observation was confirmed by Meyer-Waarden (1952) and Miyake *et. al* (1957) where the latter indicated that the peak current requirements for electrotaxis of *Kuhlia sandvicensis* decreased with the increase in length.

Since a correlation existed between the potential gradient and the density of current, the optimal potential gradients for the above reactions decreased with the increase in fish length. Explaining the above findings Meyer-Waarden (loc.cit.) stated that since large fishes receive a greater voltage in water than smaller ones they could be influenced more quickly and at relatively lower voltages. The above findings were also supported by Halsband *et. al* (1960) and Lamarque (1963). The latter considered this as a question of length of nervous elements and the projection of these nervous elements in the electric field. Since the larger fishes have greater nervous elements, when projected to external electric field, they react to potential gradients lower than those which affect the fishes of smaller length.

With regard to selectivity, the data obtained by Biswas (1977) with respect to Indian fishes and prawn clearly indicate that large animals were generally more susceptible to electric shocks than smaller ones irrespective of species and conductivity of water. This difference was due to the fact that when an electric current creates a voltage gradient in the water, a large specimen intercepted more of the gradient than a smaller one and thus is exposed to a larger potential difference. That is why a higher voltage between the electrodes was necessary in case of smaller organisms to experience optimum head to tail potential as that of larger ones.

Houston (1949) expressed that the usual commercial varieties of fish reached a state of electrotaxis when subjected to voltages between head and tail of 0.5 to 1.5 volts and electronarcosis at voltages between 1.5 to 5.0. Lamarque (1963) concluded that the fish behavior in an electric field depends essentially on the potential gradient or even better, potential difference between head and tail.

The requirement of reaction thresholds were found to be inversely proportional with the fish length in every species of animals tested. Several reasons can be put forward, but the most evident one is that larger speciemens have longer nervpos structures. Short nerves in an electric field are excited at a higher value of current than long nerves (Laugier, 1921).

In the case of *Mystus aor* fishes of 93 mm median length reacted for the first reaction, electrotaxis, electronarcosis and death at a ratio of average current densities of 1 : 2.9 : 5.7 :10.9, whereas specimens of 123 mm size responded for the identical reactions by current intensities in the ratio of 1 : 2.7 : 5.5 : 9.6. *Scatophagus argus* of 53 mm length showed the first three reactions in current intensities at the ratio of 1 : 2.1 : 3.2, while the organisms of 108 mm were influenced in current densities at a ratio of 1 : 1.2 : 1.4. However, the reaction thresholds in case of other species like *Macrobrachium rosenbergii, Labeo rohita, Cirrhinus mrigala, Notopterus notopterus, Channa punctatus* and *Wallago attu* did not follow the similar pattern. An increase in ratios between their reaction thresholds with the increase in length were invariably noticed even though the absolute current densities for the reactions were reduced with the increase in size of the animals.

Water Conductivity

Low water conductivity makes it very difficult to attain sufficient currents to produce notable responses in fish, Lennon and Parker (1958), who found extreme conductivities in Appalachian mountain streams, tackled this problem by adding salt to the water to improve electrofishing. Extremely high water conductivities call for currents too large to be supplied by portable equipment without special electrical control devices and for very high conductivity, DC is ineffective (Vincent, 1971).

Hosl (1955) described that the harder the water, the less is its resistance and the greater its conductivity. The voltage demand depends on the conductivity of the water (Hattop, 1959). Murray (1959) observed that the lower the resistivity of the water, the lower the voltage produced, and greater the current drawn. Cattley (1955) explained that in fresh water, electric currents tend to flow through the fish because of its lower resistance than the surrounding water, but in water of higher conductivity electric current tends to flow around the fish because of its higher resistance of fish body than the surrounding media. Therefore, although electric current will flow through water of higher conductivity more rapidly, more current will be required to produce similar effects in waters of higher conductivity than those of lower conductive media.

The data in respect of *Labeo rohita* and *Notopterus notopterus* clearly established the fact of requirement of higher threshold current densities for their response in the water of higher conductivity (2.8 x 1000 ohms/sq cm) as compared to that of the media of lower conductivity (9.5 x 1000 ohms/sq cm.) irrespective of fish length. The above findings corroborate those of Drimmelen (1953) where he had stated that the better the conductivity of the water, the larger the dose of electric energy necessary to elicit a response.

An increase in threshold current densities of 335 to 365.9 per cent, 291.2 to 376.9 per cent and 224.9 to 200.5 per cent for waters of higher conductivity (2.8 x 1000 ohms/sq cm over that of lower conductivity (9.5 x 1000 ohms/sq cm) were observed in case of *Labeo rohita* to respond for the first reaction, electrotaxis and electronarcosis.

In case of *Notopterus notopterus* the corresponding increase in reaction thresholds by 300 to 550 per cent, 264.9 to 270.6 per cent and 222.7 to 294.8 per cent were noticed in waters having resistance of 2.8 x 1000 ohms/sq cm., over that of the water media having a resistance of 8 x 1000 ohms/sq cm (Biswas, 1977).

The effect of electrical influence on fish is greatly dependent on the conductivity of water mass (Harris, 1953).

Salmo irideus having a median length between 146 to 196 mm exhibited first reaction, galvanotaxis and galvanonarcosis in current densities of 0.042 to 0.144; 0.156 to 0.304 and 0.571 to 0.736 δ in water having conductivity of 2 x 10000 mho/ml as against 0.067 to 0.172, 0.271 to 0.519 and 0.642 to 0.894 δ in a higher conductivity of 10 x 10000 mho/ml in impulse DC of 82/second.

Tinca tinca of 196 to 236 mm median length exhibited first reaction, galvanotaxis and galvanonarcosis in current densities of 0.204 to 0.682, 0.894 to 1.221 and 1.586 to 2.365 δ in water having conductivity of 2 x 10000 mho/ml in impulse DC of 82/second as against 0.145 to 0.154, 0.35 to 0.438 and 0.682 to 0.792 δ in 10 x 10000 mho/ml water conductivity.

Cyprinus carpio of 96 to 126 mm median length showed first reaction, galvanotaxis and galvanonarcosis in current densities of 0.037 to 0.061, 0.313 to 0.46 and 0.948 to 1.03 δ in water conductivity of 2 x 10000 mho/ml and 0.131 to 0.2, 0.48 to 0.771 and 1.35 to 1.79 δ respectively at 10 x 10000 mho/ml water conductivity in impulse DC of 82/second.

The absolute threshold current densities required for galvanotaxis and galvanonarcosis of *Salmo fario* of 86 mm rose to.963 and 1.577 δ in water conductivity of 8.2 x 10000 mho/ml from.64 and 1.57 δ respectively at 2.2 x 10000 mho/ml when subjected to impulse DC of 69/second (Biswas, 1971).

The effect of electrical influence on fish is greatly dependent on the conductivity of the water masses as well as on the body voltage and length. The conductivity of the water depends on its ionic composition and its temperature, which can vary in the extreme, as well as the conductivity of the bottom.

Ecological Characteristics

Animals of different taxonomic groups behaves differently in an electric field. Annelids respond to cathode during galvanotaxis. These differences reflect variation in the reflex activity and nervous system of the animals (Blancheteau *et al*,1961). For the worm the reflex possibility is quasi-nil, thus it moves towards the cathode, and it is only stimulated when it faces the cathode.

Differences have also been observed between species of the same animal class, particularly fish (Blancheteau *et al*, 1961; Balayev, 1981; Stewart, 1989). These differences primarily relate to the initial response, that is, a pelagic fish will swim rapidly, a benthic species will bury itself in the substrate and a cryptic species will hide itself. As the field strength increases these behavioral responses are over-riden, because the usual reflex cannot overcome the effect of the stronger electric stimulus effect.

Since the effect of electrical influence on the fish depend on the body voltage and the electrical conductivity of surrounding water mass, the electrical resistance of natural waters fluctuate due to ionic concentrations of ground waters.

Considerable fluctuations of specific resistance may occur in a short time at any water body, owing to sewage pollution, changes in temperature, falling leaves and debris etc. It is known for instance, that electric fishing gear which efficiently and easily takes eels from the Steinhuder Lake in winter and spring until the end of April, can not be used there from that date until fall. Owing to the high water temperature and the increasing reed growth, the conditions of conductivity in the water and the fish change so extremely that the energy of the electrical gear for producing an efficient electrical field is no longer sufficient.

Chapter 11

Harmful Effects of Electrical Exposure

Effects on Older Fishes

In the preceding chapters it has been shown that uncontrolled electric exposure can kill fish or produce strong fatigue. Mortality or, in less severe cases, the degree of injury depends on voltage gradient (Stewart, 1962), exposure time (Chmielewski *et. al*, 1973b; Whaley *et. al*, 1978), current form (Lamarque, 1976a), species and size of fish (Stewart, 1967; Chmielewski *et. al*, 1973b). The injuries are primarily caused by synaptic fatigue and broken bones and vertebrae (Figure 11.1).

Secondary effects, such as rupturing of the dorsal artery during vertebral fracturing and gill filament damage have also been reported (Hauck, 1949). Surprisingly, electric exposure fails to induce any of the typical stress related changes in blood lactate levels normally observed when fish are caught by other methods (Schreck *et al*, 1976, Burns and Lantz, 1978). This is because in methods other than electric fishing, fishes have to struggle for quite some time before death (asphyxia due to mass capture and confinement in a small area, as in case of purse seining or struggling for a long time after hooking in the long line), which results in breaking down of muscle glycogen into lactic acid and cause early deterioration and spoilage of fish. Electrocuted death save the fish from considerable death struggle and thus the formation lactic acid and the typical stress-related changes in blood lactate levels.

Synaptic fatigue occurs when the fish has been over-exposed to a tetanizing current. In the case, the fish does not recover immediately and dies from drowning because breathing is not re-established. Occasionally tetanus persists after the interruption of the current (*post-tetanic potentiation*) preventing the resumption of breathing.

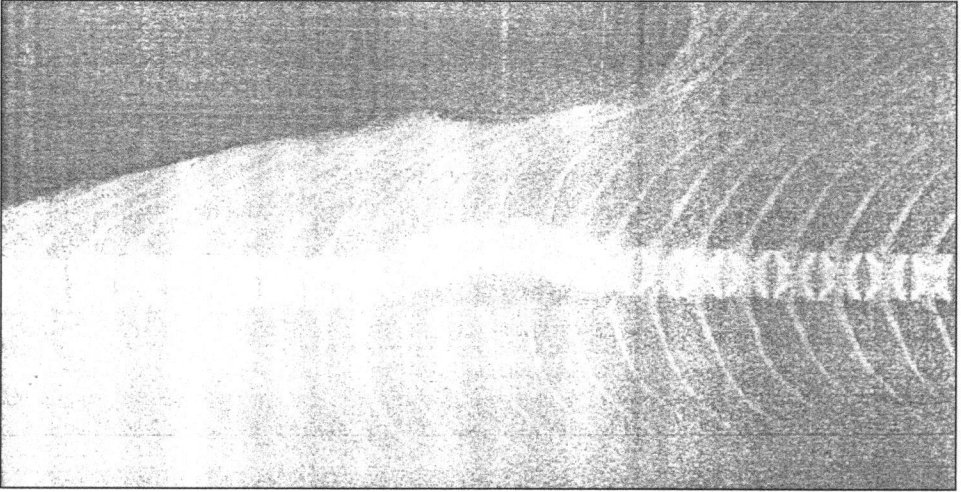

Figure 11.1: Broken vertebrae induced by square-wave current (260 V, 60 Hz, 4 ms pulse duration) used in practical electric fishing (from Sharber and Carothers, 1990).

Synaptic fatigue depends upon the species of fish and the current type. In fragile species it is important to pay attention to the current characteristics to reduce, as far as possible, the exposure time and to use an anode with a large surface area to reduce the voltage drop in the immediate vicinity of the electrode. Any fish which has been tetanized during capture should be placed in a well-oxygenated tank to recover. This is particularly important for eels. They can die from suffocation provoked by an excess of mucus on the gills produced under the action of current.

Broken bones, particularly vertebrae and bleeding are caused by violent contractions produced simultaneously by the current on both sides of the fish body following direct excitation and hyper-reflexivity. A single pulse, and sometimes a low voltage, can be sufficient to provoke this effect. With salmonids it is easy to establish whether the vertebrae are damaged through examination of the skin; the dark spots appear in the proximity of the vertebra dislocation probably resulting from the excitation of chromatophores when the sympathetic fibers are damaged. These marks are not burns. To produce them it is not necessary for the fish to come into contact with the electrode. If a large part of the body becomes dark a total rupture of the spinal column is probable. The degree of dislocation depends on the current type, voltage gradient across the fish, fish species, the physiological state of the fish (namely, decalcification produced by spawning; Stewart, 1967) and poor food (with a probable lack of magnesium and calcium).

Several studies have been carried out to quantify the harmful effects of different electric currents on fish. Pratt (1955) and Spencer (1967) showed that exposure to AC caused considerably greater mortality than pure DC. Lamarque (1967a) suggested that pulsed DC and AC can provoke violent tetanus and, depending on the exposure time, paralysis may follow due to synaptic exhaustion. Mortality tests carried out by Lamarque (1967b) showed that heaviest loses (more than 80 per cent) occurred with condenser discharges (80 Hz) and rectified AC (90 Hz). 500 volt DC produced relatively low mortality (less than 10 per cent).

The most commonly referred damage caused by electric fishing involves vertebral malformations (Hauck, 1949; Klein, 1967; McCrimmon and Bigwood, 1965; Spencer, 1967). Stewart (1967) showed that these injuries are of two types; those caused by DC are compacted (compression of several vertebrae) whereas with AC they are of oblique type (misalignment of successive vertebrae) In their recent and succinct study, Sharber and Carothers (1989) found that when pulsed DC was used approximately 50 per cent of fish suffered spinal injuries, involving an average of eight vertebrae. They also observed that the number of fish injured was a function of the pulse shape, with exponential pulses and square-wave (44 per cent injured) being less damaging than quarter-sine wave (67 per cent).

Percentage Mortality Resulting from Rainbow Trout being Exposed to different Electric Current Types for 20 seconds at 20 cm and 50 cm from the Anode

Current Characteristics	Percentage of Mortality	
	50 cm from Anode	20 cm from Anode
Direct current, 500 V	0	6
Smooth, half-wave rectified AC, single-phase, 400 V	0	17
Condenser discharges, 50 per cent duty cycle	0	86
Condenser discharges, 33 per cent duty cycle	0	93
Square-waves 66 ms, 5 Hz, 400 V	0	50
Rectified AC, single-phase, 90Hz, 400 V	27	89

In general, it appears that DC is the least and AC the most harmful electrical output, with pulsed DC falling between the two. A clarification for these differences is provided by the neurophysiological and behavior responses of fish to the different currents.

Usually fish is caught when more or less facing the anode and this status is important. When a fish faces the anode in DC, the inhibition on the motor pathways protects the fish against hyper-reflexivity (*anodic protection*). It can remain in this position, at the current value for galvanonarcosis, for several hours without any trouble (Lamarque, 1976b). Nevertheless damage can occur when the fish lays motionless and tetanized in the vicinity of the cathode or when the current is abruptly re-established. In the later case the current is acting as pulsed DC at a high voltage (In electric fishing, DC is always operated at a higher voltage than pulsed DC because of the higher threshold for a particular response).

With pulsed DC, the extent of injury depends mainly on pulse duration and frequency. The most harmful currents are those with a pulse duration of 2-5 ms at 5-200 Hz. The normal swimming mechanism is a sinusoidal curvature of the body with the muscles on one side of the body contracting, whilst those on the other side are relaxed (Nursall, 1956). When the fish are stimulated by such pulsed currents, the muscles on both sides of the body are excited at the same time submitting the vertebral column to opposing constraints which break or dislocate the vertebrae.

Amongst the pulsed the currents, condenser discharges are the most harmful (Lamarque, 1967a). It is possible to kill an eel in 30 seconds with this current. These discharge pulses have a steep initiation slope and short pulse duration which allows little anodic protection. Unfortunately, they have a great neurophysiological impact (Halsband, 1967) and are frequently used in electric fishing. It is wrong to believe that the best results in electric fishing are achieved by a high voltage or tetanizing current which is often chosen for such peculiarities. Low frequency (less than 50 Hz), short duration square-wave discharges are similar in effect. About the same effect is found with low frequency (less than 50 Hz) square-wave pulses. For example, with a square-wave current of 50 Hz, 2 ms duration (duty cycle of 10 per cent) the pulse duration is superior to the useful time and a closing reaction occurs with each pulse. After the pulse there is a time without current of 18 ms leaving time for the nerve impulse to act fully on the muscles. This type of current is particularly useful for producing a general tetanus, as the pulse acts on the whole system.

Rectified AC offers different degrees of protection depending on the frequency and mode of rectification (In case the current can be smoothed by adding a high value condenser in parallel with the output. If the current is sufficiently smoothed no damage will occur, but such a current is not easy to use under continually changing electric fishing operating conditions) For example, half-wave, single-phase rectification at 50 Hz produces a 10 ms stimulation followed by a similar time without current which allows hyper-reflexivity of the muscles without anodic protection. This current is harmful through excessive tetanus (and does not save energy). Quarter-sine wave currents (5-60 Hz) derived from half wave rectification has a pulse duration of 5 ms and off time of 15 ms and exposes the fish to the full effects of the current without protection. Full-wave rectified current at 50 Hz is continuous and therefore offers some protection. At 300 Hz, half-wave rectification has only a short period with current (1.5 ms) and is less damaging than 60 Hz. Full-wave rectified AC at original AC frequency of 300 Hz will have an effective frequency of 600 Hz thus producing a continuous current similar to, and acting like, pure DC. Three-phase rectified AC is potentially the best available because the voltage, although undulating, never drops to zero, thus it is similar to DC and affords anodic protection (Lamarque, 1976a). The undulation rate for half-wave rectified current is 17 per cent but only 4 per cent for the full-wave rectification. These two currents produce about the same effect as DC but the full-wave is less tetanizing (Lamarque, 1976a). These transformed currents can thus have great advantage when compared with DC produced from a transformer (Lamarque *et.al*, 1978).

Alternating current is the least satisfactory for electric fishing from the point of view of injury; it offers no anodic protection and induces maximum tetanizing effect. It seems that AC at low frequency (= 50 Hz) is more damaging than higher frequencies (more than 300 Hz).

Effects on Egg and Juvenile Fishes

Scheminsky (1922) studied the effect of trout eggs exposed to DC for long periods. He observed a shifting of the embryo in the egg and, after long exposure, a high mortality. However the experimental conditions were far from those that exist in electric fishing practice. Godfrey (1957) found that eggs in the early cleavage stage (that is, when the first cellular differentiation takes place) are more susceptible than pre-cleavage (zygote) eggs. Maxfield *et. al* (1971) showed that uneyed eggs were more vulnerable than eyed eggs in salmon and rainbow trout. Thus it appears that eggs are most vulnerable between fertilization and the eyed stage and electric fishing over reeds or other spawning areas should be avoided when fish are at this developmental stage.

Juvenile fish such as fry and alevins tend to be extremely fragile and mortality can occur due to handling stress as much as the current. Mortality therefore tends to depend on the species and the physiological state of the fish as well as current characteristics. For example, mortality is very common with pikeperch (*Stenzostedion lucioperca* L.*)* but rare with trout fry. A salmon parr will not suffer unduly from a current that kills the older smolt. In general, however, the current characteristics that are dangerous to juvenile fish are similar to those described for adults.

Effect of Electric Exposure on the Developing Fish Eggs and Embryos of Indian Origin

The effects of external interference on the embryonic development of fish eggs has received only sporadic study during the last century. Amongst those, the effect of environmental conditions upon development has received attention. Temperature changes during incubation dislocate the developmental process in an observable way (Needham, 1942; Makino and Osima, 1943; Thomopoulos, 1954; Lieder, 1953; Lindsey, 1954; Hayes *et. al*, 1953). The inhibitory effect of light and the lethal effect of ultraviolet light on the developing eggs of salmonid fish were studied by Hinrichs, 1925; Haempel and Lechler, 1931 and Stuart, 1953. High mortality in the eggs of the Pacific herring due to high concentration of carbon-dioxide was observed by Kelley (1946). The eggs and larvae of the same species have been found to tolerate a wide range of salinity change without any effect on the rate of development (Mc Mynn and Hoar, 1953).

Very little information is available on the effect of the experimentally induced interference on the fish eggs, except for a note on the differential sensitivity of *Fundulus* eggs to graded dose of X-rays (Rugh, 1954) and the effect of DC shocks of different severities to trout and salmon ova in various stages of development (Godfrey, 1957). Practically no information is available on the effects of electrical irradiation on the development of Indian fish eggs.

Exposing the fertilized carp eggs to different degrees of electrical flux were carried out in order to find out the effect on the development and hatching of the embryos and also to determine the lethal limits of exposure, so as to establish precautionary measures in order to prevent egg damage when applying similar electrical fields, in waters of similar conductivity, in the vicinity of waters containing live eggs. (Biswas, 1977).

Fertilized eggs of Indian major carps (*Labeo rohita, Catla catla, Cirrhinus mrigala* and *Labeo fimbriatus*) in their early cleavage (5-6 hours of incubation), late cleavage (8-9 hours of incubation) and near to hatching (10-12 hours of incubation) were exposed to underwater electrical field of varying intensities for different time periods.

The exposure to the developing eggs in the electrical field (AC or DC) was done by (a) direct stimulation where individual egg came in contact with the stimulating electrodes and (b) indirect stimulation where the eggs were exposed to electric flux through the fresh water media. The developing eggs were treated with short electric exposure for 0.5 to 360 seconds. The tests were carried out in the normal rearing media for the developing carp eggs at an optimum temperature for the development of fertilized carp eggs (25 to 30.5 degree Celsius) The electrical resistance of the media varied from 9000 to 50000 ohms per sq. cm. Current density of 0.0112 to 0.2884 uA/sq. mm was established for every volt of applied potential depending on the resistance of the media.

Visible Effects during Shock Treatment

0 Stage (Normal)

The principal characteristics of the developing eggs under normal conditions, in all the species, from an early cleavage stage (that is, 5-6 hours after fertilization) are the rate of further development till they hatch out. The time required for attaining a coma shaped embryo and the formation of notochord (late stage development, 8-9 hours after fertilization); onset of twitching movement of embryo within the egg shell, appearance of eye spot, hatching out of embryo by breaking the egg shell and the survival of hatchlings have been presented for each of the four species before giving any current exposure to them.

In the case of *Labeo rohita*, the developing embryo attains a coma shape within 8 to 9 hours, the twitching within the egg shell was observed between 10 to 11 hours and the final hatching of the embryo (by breaking the egg shell) was observed within 15 to 16 hours, all, after fertilization. The survival percentage was 82 to 87 when reared in river water, pond water or non-chlorinated tap water. In filtered water however, the survival rate varied between 13.3 to 21.0 per cent.

The embryo of *Cirrhinus mrigala* under normal conditions attained a coma shape within 10 to 11 hours, after fertilization. They started the twitching movement within 12 to 13 hours and hatched out within 16 to 17 hours after fertilization. 94 to 97 per cent of survival of the hatchlings were observed in pond water, river water and non-chlorinated tap water, but survival rate was 28.2 to 34.0 per cent in filtered water.

Catla catla embryos under similar conditions attained a coma shape within 7 to 8 hours, and showed twitching movements with in 9 to 10 hours. They hatched after 14 to 15 hours of fertilization. 82 to 94 per cent hatchlings survived when reared in river water or pond water or non-chlorinated tap water. In filtered water, however, the survival rate was 8.2 to 11.4 per cent.

The embryos of *Labeo fimbriata* developed in to a coma shaped in 11 to 12 hours after fertilization. They started twitching movements within 13 to 14 hours and hatched within 18-19 hours after fertilization. 95 to 98 per cent of the hatchlings survived in river water or pond water or in non-chlorinated tap water. In filtered water only 31.4 to 41.8 per cent of embryos survived.

The visible effect of a DC field on the developing eggs of *L. fimbriata* and *Cirrhinus mrigala* were classified according to the intensity and period of treatment.

Stage I (During DC Shock Treatment)

No visible effect was observed during the shock treatment in the developing eggs of *L. fimbriatus* when treated during early or late cleavage or near to the hatching conditions. The twitching movements of the embryos continued within the egg shell during the passage of current in all the media.

C. mrigala eggs when treated in their early cleavage conditions, elongated and burst in 6.8 to 21.4 per cent of the cases depending on the period of treatment (60 to 360 seconds). In the late cleavage conditions, 3.1 to 5.6 per cent of developing embryos burst when exposed to electric flux for 60 to 360 seconds, under similar conditions.(Figure 11.2) However, the twitching movement of the embryos within the egg shell were not affected during the shock treatment.

Stage II (After Shock Treatment)

The residual effect of shock treatment on the developing embryo of *Labeo fimbriata* in early and late cleavage conditions were characterized by;

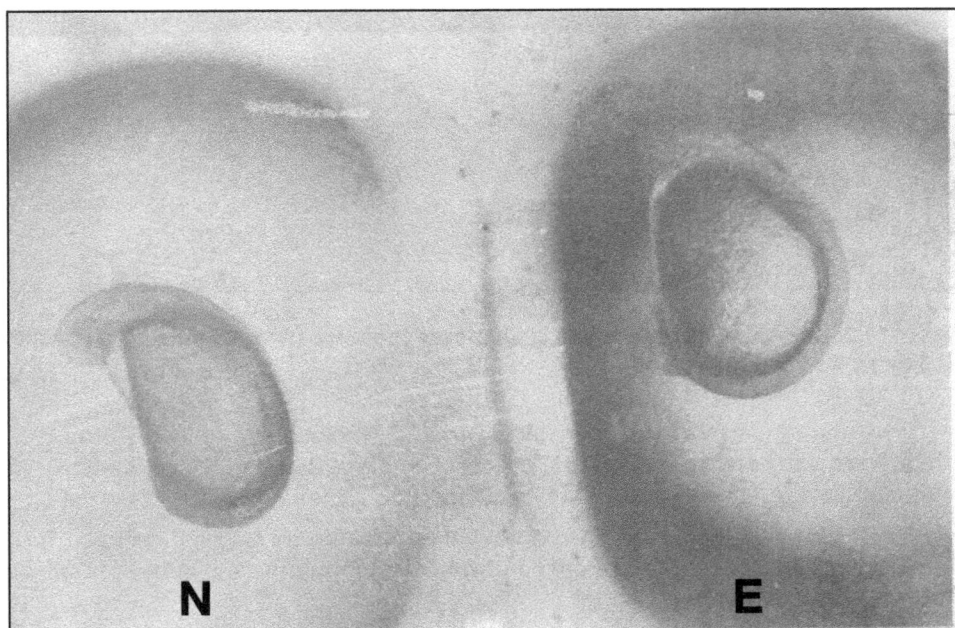

Figure 11.2: Effect of DC shocks on the developing eggs if *Cirrhinus mrigala* Bursting of embryo (Magnification x 10.5). N: Normal; E: Experimental.

(a) the bursting of embryo,

(b) development of white spots on the yolk,

(c) the abortion of the embryo from the egg shell.

This varied within one hour after the shock from 1.6 to 17.1 per cent, 10.8 to 17.3 per cent and 0 to 3.2 per cent respectively and depended on the period of treatment. After two hours of treatment, 18.9 to 30.1 per cent of the embryos showed twitching movement. The hatching of the embryos occurred in 70.2 to 76.1 per cent of which 59.0 to 57.1 per cent developed in normal conditions when the treatment was for 180 to 60 seconds; while in 360 seconds of exposure 57.0 per cent of embryos developed in to abnormal condition (Figure 11.3).

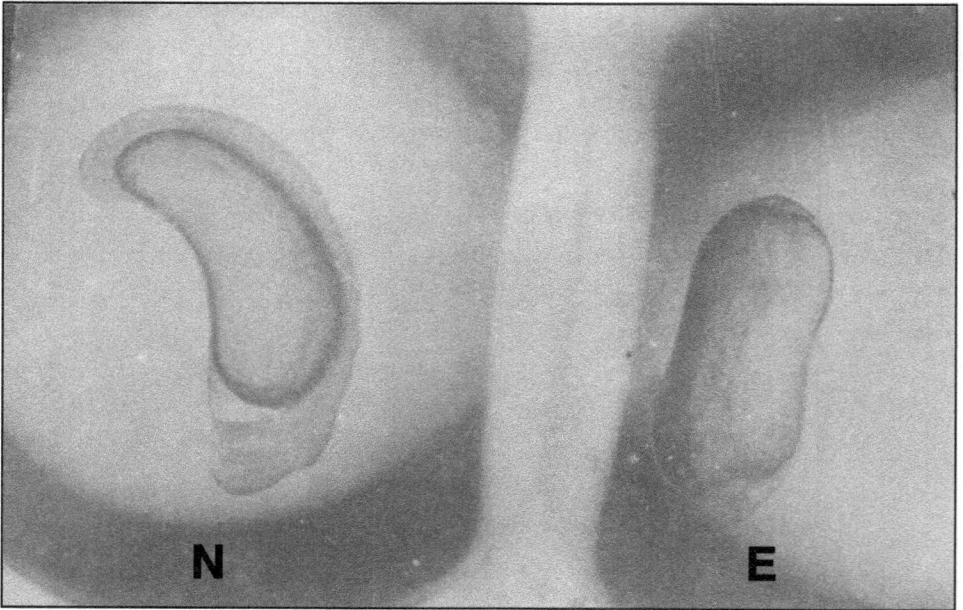

Figure 11.3: Development of embryos in abnormal condition (After 6 hours of treatment, 10.5 x). N: Normal; E: Treated.

Two hours after shock treatment there was a cessation of development in embryos which were near to hatching. The percentage of embryos that showed appearances of white spots varied from 5.8 to 35.0 respectively.

These embryos, however, hatched out after eight hours of stoppage of current flow (40.0 to 84.2 per cent) and showed normal development even after 48 hours of stoppage of shock treatment.

After four hours of shock treatment in the early cleavage condition 13.3 to 47.8 per cent embryos of *Cirrhinus mrigala* turned opaque where further development stopped, this being subject to the period of treatment. Eight hours after shock treatment,

43.2 to 81.8 per cent of the embryos showed twitching movements within the egg shell. 6.9 to 27.2 per cent of embryos were found still within the egg shell, thirteen hours after shock treatment, when these were treated for 360 to 60 seconds. Total hatching under this condition were 40.3 to 72.7 per cent of which 31.8 to 72.7 per cent developed normally and 9.5 to 0 per cent survived but in a deformed condition (Figure 11.4).

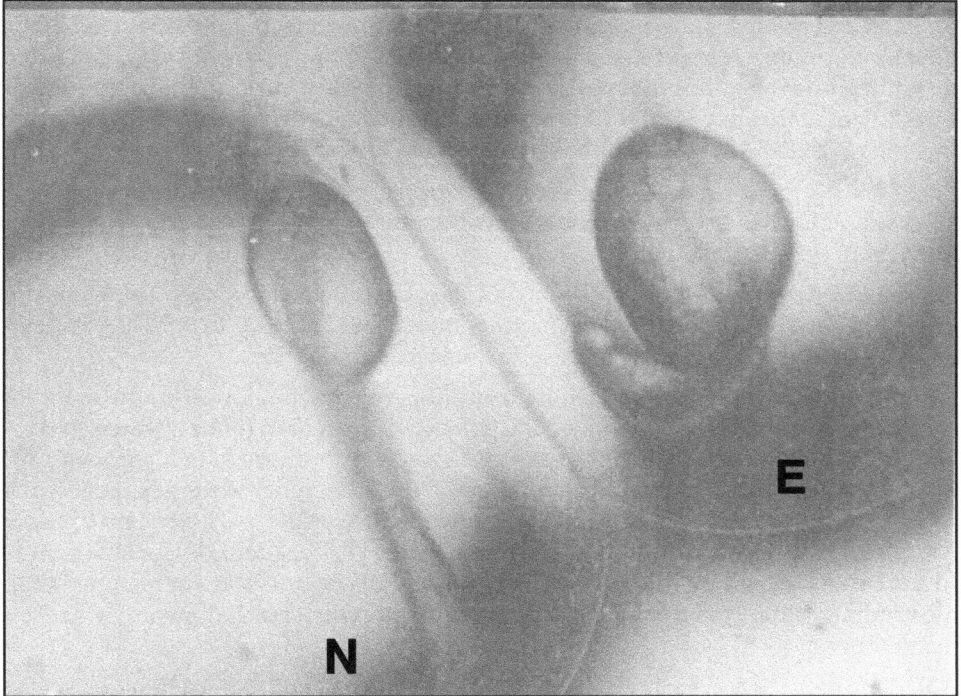

Figure 11.4: Deformed embryo (After 3 hours of treatment; 10.5 x). N: Normal; E: Treated.

When treated in the late cleavage condition, 5.2 to 0 per cent of the embryos stopped developing two hours after stoppage of treatment; 52.4 to 100 per cent of them showed movements within the egg shell after six hours of treatment and were found still within the egg shell even after 13 hours of treatment, when treated for 360 to 60 seconds. The survival rate of the embryos ranged from 32.1 to 100 per cent of which 0 to 98 per cent developed normally and 32.1 to 2.0 per cent developed abnormally.

Even seven hours after shock treatment 28.3 to 44.7 per cent of the embryos remained in the egg shell even though they were in a near to hatching condition. The hatching percentages of embryos were found to be 43.5 to 100.0, of which 8.0 to 95.0 per cent showed normal development and 35.5 to 5.0 per cent survived in deformed condition. These were exposed to electric shocks for 360 to 60 seconds.

Stage – I (During AC Shock Treatment)

When exposed to AC field in filtered water, the embryos in early cleavage stages of *Labeo rohita* elongate and burst into fragments (during passage of current), the percentage of which varied from 4 to 28 depending on the intensity and period of shocks. In other media, like river water, pond water or non-chlorinated tap water no visible effect was observed during the shock treatment.

Treated in the late cleavage conditions, no visible effects were observed in the developing embryos in filtered water. In river water, pond water or non-chlorinated tap water, however, the embryos developed opaque rings around their extremities. Bursting of embroy around the extremities of embryos was noticed in high current intensity (1.228-1.421 µA/sq mm) in 100 per cent cases.

Twitching movements of the embryos stopped during the passage of current in all cases. After the stoppage current flow the embryos were found to regain their movement one after another.

Cirrhinus mrigala eggs in their early cleavage and late cleavage conditions did not exhibit any visible effect when exposed to AC field of 0.307 to 0.726 µA/sq mm even up to 180 seconds in any of the afore mentiomed media.

When treated in near to hatching condition, the movements of all the embryos within the egg shell stopped during the passage of current in filtered water. In river water, however, the movement of the embryos within the egg shell stopped only when their body axis was parallel to field lines. When the embryos were perpendicular to current lines, their rate of twitching movement was 26 to 31 per cent and not completely stopped during the current exposure. 77.5 per cent of egg shells were found broken expelling the embryos, where only 10 per cent of them regained their movements leaving 90 per cent dead, the body of which curved inwardly at their posterior end.

Catla catla eggs in early cleavage and late cleavage conditions when exposed to current intensity of 0.614 µA/sq mm in pond water did not exhibit any visible effect during the passage of current. But near to hatching condition the movements of all the embryos remained suspended during the current flow without their being expelled from the egg shell, The embryos regained their movement one after another on switching off the current.

The developing embryos of *Labeo fimbriatus* in early and late cleavage, when treated in current density of 0.726 µA/sq mm in river water media, did not exhibit any visible effect during shock treatment of 180 seconds. On further treatment for 360 seconds, however,100 per cent of the embryos were dead, ruptured along their extremities (Figure 11.5). The effect on the embryos in near to hatching conditions, were similar to those in the case of *Catla catla*.

Stage – II (After Shock Treatment)

The after effect of shock treatment to the fertilized eggs of *L. rohita, C. mrigala, C. catla* and *L. fimbriatus* when activated in their early cleavage, late cleavage and near to hatching condition by AC field to varied degree of intensity and period of exposure

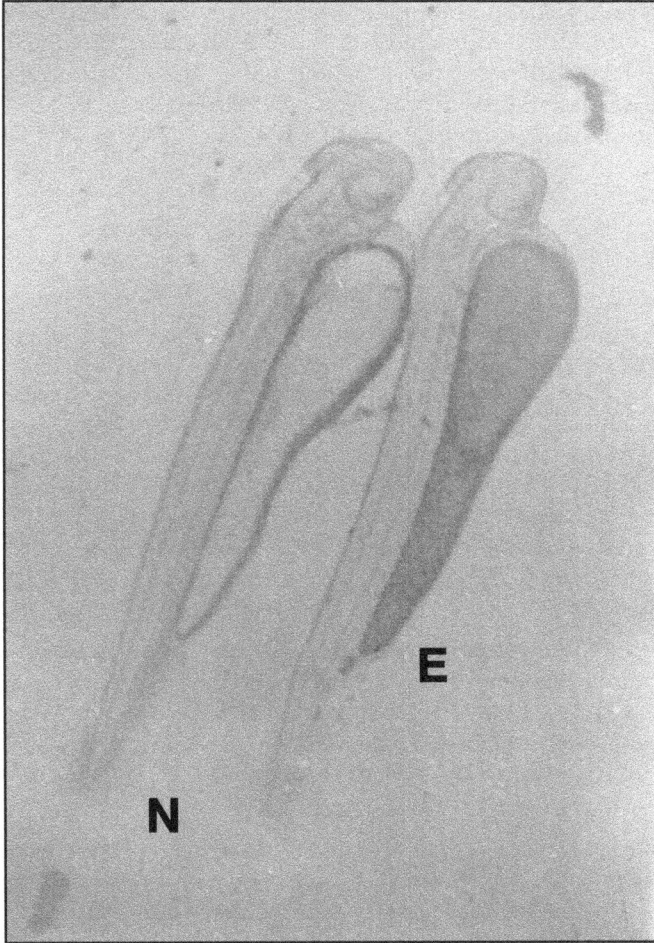

Figure 11.5: Rupture of embryo along the extremiyies (10.5 x). N: Normal; e: Treated.

indicated the survival rate of 0.6 per cent of *L. rohita* eggs in normal condition, when exposed to a current density of 0.304 µA/sq mm for 180 seconds in near to hatching conditions. 10.3 per cent of *C. catla* eggs in near to hatching condition developed normally when treated with a current intensity of 0.614 µA/sq mm for 15 seconds. The embryos of *C. mrigala*, however, developed normally in 66 to 100 per cent and 86.5 to 100 per cent of the cases when treated for 60 to 180 seconds in a current density of 0.307 to 0.726 µA/sq mm in their late cleavage and near to hatching conditions respectively.

Labeo fimbriatus eggs in their late cleavage and near to hatching condition when exposed to current density of 0.726 µA/sq mm for 60 to 360 seconds, showed normal development in 0 to 78.1 per cent and 34 to 84.4 per cent of the cases respectively. The rest of the eggs, after shock treatment, died in the course of time on account of premature breaking of the egg shell, bursting of embryos and abnormal development of various degrees. The normal development of all the eggs in early cleavage, irrespective of species, ceased by way of premature breaking of the egg shall, rupture of the embryos and abnormal development when subjected to electrical stimulation of 0.231 to 0.726 µA/sq mm for 15 to 260 seconds. Premature breaking of the egg shell subsequent to electrical treatment was more prevalent in late cleavage and near to hatching condition (31.4 to 100 per cent), while the bursting of embryos under the similar condition was predominant in early cleavage condition (33.3 to 100 per cent) in all the species studied (Biswas, 1977).

On a comparative basis, the hatching of *L. fimbriatus* eggs after shock treatment of 0.726 µA/sq mm in late cleavage stage was more (57.6 to 88.8 per cent) in DC field than (0 to 87.2 per cent) in AC field. Similarly, eggs of *C. mrigala* when treated in their cleavage, late cleavage and near to hatching condition showed 40.9 to 72.7 per cent ; 32.1 to 100 per cent and 43.5 to 100 per cent hatching in DC field as against 0 to 26.2 per cent ; 0 to 31.4 per cent and 10 to 74.8 per cent in AC field. 360 seconds of exposure in AC field was found to be fatal in 100 per cent cases for early cleavage and late cleavage condition irrespective of the species. Even 60 seconds exposure in AC field in early cleavage and late cleavage condition resulted 9.1 to 13.4 per cent of deformed embryos in both the species as against 2 per cent deformities in DC field in case of *C. mrigala* only.

The eggs of *C. mrigala* were exposed to electric shocks directly by touching the electrodes on the egg shells and also through water media. In either case, maintaining a terminal voltage of 8.5 and varying the period of treatment from 2 to 60 seconds, in both direct and alternating current, the percentage of hatching was 74.8 to 100 per cent, when treated through fluid media as against 0 to 20 per cent hatching and survival during direct contact with the eggs.

As regards to different doses of current intensities on the developing eggs of *L. rohita* in non-chlorinated tap water, the cleavage of all the embryos stopped after shock treatment when the eggs in early cleavage and late cleavage conditions were exposed to current densities of 17.04 µA/sq mm; 9.088 µA/sq mm; 3.408 µA/sq mm and 2.499 µA/sq mm for five seconds. When the eggs in near to hatching condition were treated in the above mentioned current intensities for a similar period, only 5 per cent and 7 per cent of the embryos hatched out with normal development in current intensities of 3.408 µA/sq mm and 2.499 µA/sq mm respectively as against 82 to 87 per cent survival under controlled conditions without any shock treatment.

The effect on *L. rohita* eggs, when treated in river water (16000 ohms/sq cm) in their different stages of development, showed 85.2 per cent, 62.5 to 65.6 per cent and 80.3 to 95.6 per cent hatching of the embryos when they were treated to current intensities of 4.807 µA/sq mm; 4.807 to 3.077 µA/sq mm and 7.692 to 5.128 µA/sq mm for 180 seconds. After 48 hours of shock treatment, 81 per cent, 60.8 to38.1 per cent and 62.4 to 91.2 per cent of these hatchlings survived in normal condition and 4.2 per cent, 1.7 to 26.5 per cent and 17.9 to 4.4 per cent in deformed condition.

The eggs of the same species when treated for 60 seconds in filtered water media (50000 ohms/sq cm) exhibited 6 per cent, 2 to 17.5 per cent and 34.4 to 92 per cent hatching and normal development in early cleavage, late cleavage and near to hatching stages under the influence of current intensities of 0.304 μA/sq mm; 0.632 to 0.304 μA/sq mm and 0.708 to 0.304 μA/sq mm respectively.

The effect of duration of electrical exposure in a field intensity of 2.499 μA/sq mm, revealed that in early cleavage condition, the percentage of hatching increased from 2 to 35 with shock duration of 15 to 0.5 seconds. But the normal development of the embryos was noticed only during 0.5 to 1 second treatment. In late cleavage and near to hatching stages, however, 100 per cent hatching with normal development were observed when treated for 0.5 to 2 seconds. During 5 to 15 seconds of current exposure 45 to 8 per cent and 72 to 44 per cent of hatching took place in late cleavage and near to hatching conditions. Normal development of the embryos continued in all the cases except in those which were exposed to current for 10 to 15 seconds in late cleavage and 15 seconds in near to hatching conditions.

Hundred per cent mortality of *L. rohita* eggs were observed within a hour of 60 seconds shock treatment (current density 2.499 μA/sq mm) in all the three developmental conditions. In 5 to 15 seconds of shock treatment, the percentage of mortality increased in different degrees with the passage of time depending on the period of shock treatment. 100 per cent of them died within 48 hours of treatment except in 5 seconds exposure to near to hatching embryos, when the mortality did not exceed 95 per cent even after 48 hours of treatment.

Treated in current intensity of 17.04 μA/sq mm for 5 seconds in early cleavage and late cleavage conditions, 100 per cent mortality within a hour observed. Lowering the current densities to 3.408 to 9.088 μA/sq mm, the mortality of embryos in all stages increased gradually with the course of time and 100 per cent mortality were noticed between 6 to 48 hours after shock treatment, with the exception of 2 per cent survival in near to hatching condition (the current intensity of the field being 3.408 μA/sq mm).

In an AC field the incubation periods of *L. rohita* eggs were 22, 21 to 16 and 0.5 hours when the treatment were given to the eggs in early cleavage, late cleavage and near to hatching conditions respectively with shocks of 3.884 to 7.692 μA/sq mm for 180 to 360 seconds as against 7 to 8 hours of incubation in case of untreated eggs.

Labeo fimbriatus eggs, when treated in DC field of 0.726 uA/sqmm in their early cleavage and near to hatching conditions for 60 to 360 seconds, hatched out between 8.5 to 9 and 8 hours respectively, as against 14 to 15 hours of incubation in case of untreated ones.

The incubation periods of *C. mrigala* eggs, when treated in their early and late cleavage stages in an identical way as that of *L. fimbriatus* eggs, were 13 to 13.5 and 11 to 11.5 hours respectively, whereas the untreated embryos hatched out within 12 to 13 hours.

The deformities of the hatchlings, who have survived from the shock treatment both in DC and AC field, includes tail defects of *L. fimbriatus* embryos in 2.1 per cent of the cases when the eggs were treated in early cleavage condition for 180 seconds at an intensity of 0.726 μA/sq mm in DC field (Figure 11.6).

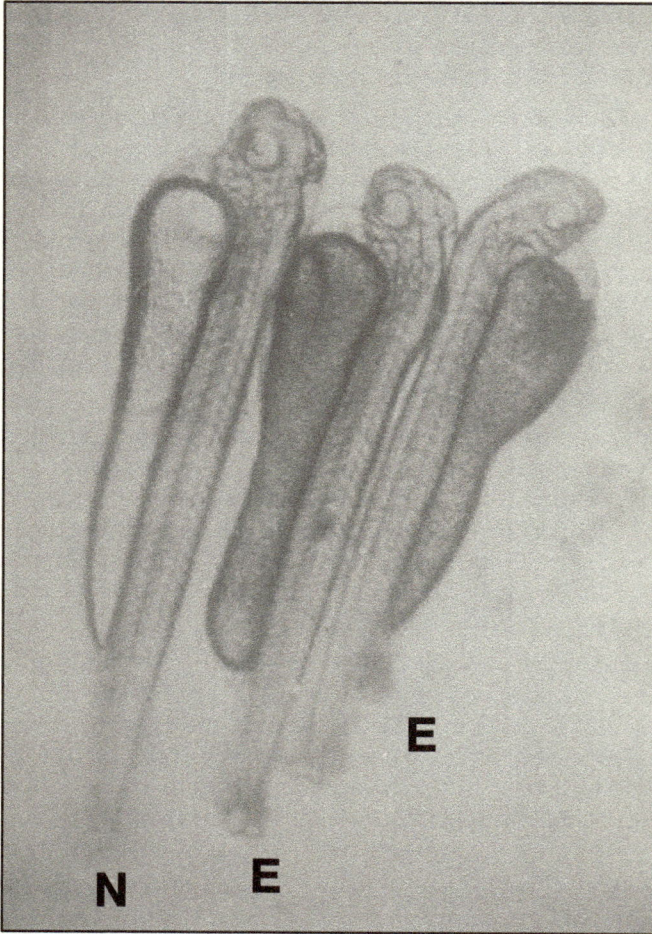

Figure 11.6: Tail defects of *Labeo fimbriatus* embryos in DC shocks (15 hours after treatment, 10.5 x). N: Normal; E: Treated.

In AC field, however, bending and an undeveloped notochord was noticed in 5.0 to 13.5 per cent and 2.0 to 7.1 per cent *C. mrigala* and *L. rohita* hatchlings when treated for a period of 60 to 360 seconds at an intensity of 0.304 to 7.692 µA/sq mm in different developmental stages (Figure 11.7).

The effects of shock treatment on the fertilized and developing fish (carps) eggs, during and after shock treatment, described in earlier paragraphs were mainly developmental and phenotype irregularities.

According to the theory of electrical conduction, the current density of the media per unit potential difference applied, increased with the conductivity of the media, which depend on the available ionic concentration of the media. Since the stimulating

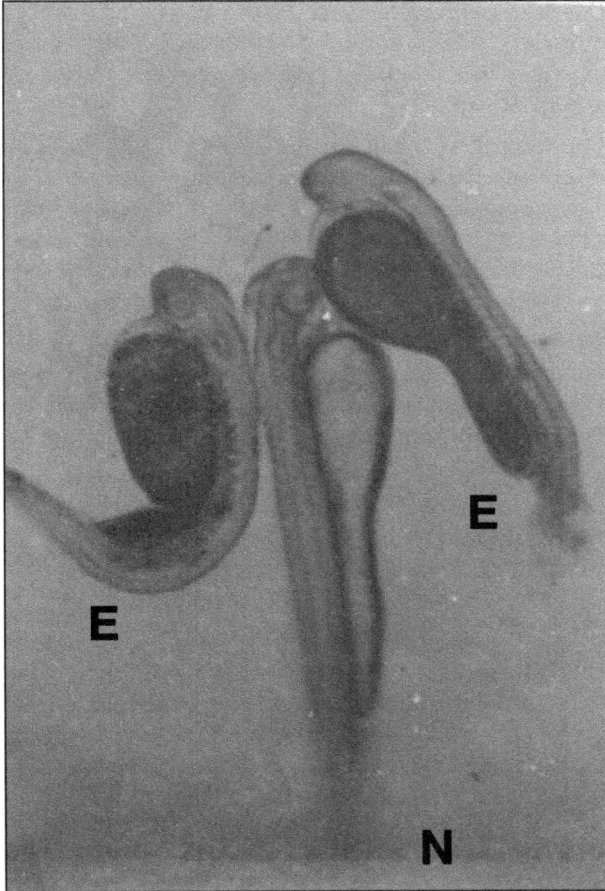

Figure 11.7: Bend and undeveloped notochord of *Labeo rohita* **embryos in AC shocks (15 hours after treatment; 10.5 x). N: Normal; E: Treated.**

and rearing media, even for the untreated ones,were kept uniform, the effect of shock treatment could be clearly isolated without it being superimposed by other factors.

The rupture of embryos in early and late cleavage conditions, premature breaking of egg shell in near to hatching stage and stoppage of twitching movement of the embryo within the egg shell during AC shock indicate its severity over DC shocks.

The residual effect from AC shocks resulted 100 per cent mortality in early cleavage stage in all the above varieties indicating the eggs in early cleavage condition are most susceptible to AC shock.

The direct stimulation of the egg brought about severe effect of stoppage of development and insignificant hatching as against 74.8 to 100 per cent hatching when treated through fluid media where the embryos received a small amount of current, being protected by the resistance of the surrounding media.

Increasing intensity of stimulus and the duration of shocks is associated with mortalities of embryos. The rate of embryonic mortalities after shock treatment also depends on the severity of the shock and possibly due to its phenotypic abnormalities produce fewer abnormal adults.

The incubation period for the untreated eggs in identical conditions was 14-17 hours for all the species studied (15-18 hours at a temperature of 27 degree Celsius in respect of Indian carp eggs, according to Chaudhuri, 1969) The mean hatching time of *L. rohita* eggs after shock treatment in late cleavage conditions in an AC field was 18.5 plus minus 2.5 hours as against 7.5 plus minus 0.5 hours for the untreated ones, showing, that the development of embryos slowed down as a result of electrical stimulation.

L. fimbriatus eggs, however, when stimulated in early cleavage and near to hatching conditions, with different severity, showed a mean hatching period of 8.5 plus minus 0.5 hours as compared to 14.5 plus minus 0.5 hours for the untreated ones.

The mean hatching period of *C. mrigala* eggs, when treated in a DC field of identical severity as compared to that of *L. fimbriatus*, was 12.0 plus minus 1.5 hours as against 12.5 plus minus 0.5 hours required for the controlled ones, indicating that the exposure to DC current causes abortion of the embryo.

Among the phenotypic abnormalities of the hatchlings from those who survived the shock treatment, tail defects were noticed in DC field (2.1 per cent) in *L. fimbriatus* embryos; while defects in the notochord (bend and undeveloped) were prevalent in AC field (2.0 to 13.5 per cent) in case of *L. rohita* and *C. mrigala* embryos proving once more the severity of AC shocks over DC (Biswas, 1977), similar to that of ultraviolet light treatment to *Fundulus* eggs (Hinrichs, 1925).

Harmful Effects of Electric Shocks on Human Beings

Significance of Electrical Accidents

The human body is adversely affected by electric current which passes through it. The consequences of this effect depend on the amperage, the kind of current (direct current, alternating current, high frequency current, pulsating current), the time of exposure and the way the current passes through the body as well as the individual resistance.

According to Koeppen the following degrees of amperage are to be distinguished.

Degree I upto 25 mA

☆ 0.1 to 1 mA-slight contractions of the muscles in the fingers

☆ 0.8 to 2.4 mA- concussion of the nerves in the fingers up to the endoderm.

☆ 9 to 15 mA- releasing the contact still possible.

☆ 9 to 22 mA – releasing the contact impossible without help.

☆ Increase of blood pressure depending on the amperage. No influence on the beating of the heart and the central nervous system.

Degree II upto 25 to 80 mA

☆ 28 to 50 mA still bearable amperage, without unconsiousness setting in.

☆ Irregular beating of the heart, increase of blood pressure, reversible stand still of the heart.

Degree III – More than 80 mA

Fluttering of the ventricles of the heart

Degree IV – 3 to 8 A

Pulmonic palsy – Increase of blood pressure, standstill of the heart, irregular rhythms

The intensity (I) of current passing through the human body depends on the voltage existing at the moment of contact (Ua) hereafter reffered to as contact-voltage and the electric resistance (Rk) of the body. Accordind to Ohm's law, the formula is;

$$I = Ua/Rk$$

The contact voltage is the amount of voltage which can be endured by the human body. It can be measured, for instance, between head and foot or between left and right hand respectively according to the parts of the body where the current enters and leaves. Water or earth as well can be current conductors. The electrical resistance of the body fluctuates considerably. It consists of the resistance of the skin and the internal resistance of the body which depend greatly on temperature, voltage, period of influence, the points of contact on the body, pressure of the contacts on the body, condition of the skin and various physiological and psychological factors. Wet skin, represents the most unfavorable case as the total resistance then amounts to the mere internal resistance of the body, which has been proved to be about 800. Fifty mA is the dangerous maximum amperage which means that the dangerous contact voltage is;

$$800 \times 0.05 \text{ A} = 40 \text{ volt}$$

If this voltage is exceeded, the lives of human beings will be in danger. To be on the safe side, the maximum contact voltage should not exceed 24 volts and the Association of German Electro-Engineers have recommended this.

Experiences showed that direct current is not so dangerous as alternating or impulse current of 40 to 600 cycles and that high frequency currents of millions of cycles are not perilous (*e.g.* diathermic frequency). The effect of Faradic stimulation on the body can still be felt up to 500 Kc.

Currents of more than 50 mA are dangerous to life if the influence exceeds of 0.2 seconds, whereas brief shocks apparently have no damaging effects on health. Finally the manner in which the current runs through the body is of importance. If the current entered the index finger and left the thumb of the same hand, only burns might occur, but deadly fluttering of the ventricles of the heart and paralysis of the breathing organs must be expected if the heart and the central nervous system are in the circuit and amperage and period of influence are sufficient. This is the case when the current for instance, runs from one hand to the other or from hand to foot or from one foot to the other. In the later case the amperage has to be double the amount to show the same effect.

Particular Dangers of Electrical Fishing

Accidents are always possible because of the kind of current and voltage used for fishing. This applies to direct current as well as to impulse current. Because of the wet hands and feet, the resistance of the skin is especially small.

The following accident possibilities exist.

In freshwater fishery with two electrodes (cathode and anode), the worst case would be if each hand touched one of the electrodes at the same time. The human body would feel the total voltage, the heart would be in the circuit and a deadly effect would be unavoidable if the current continued to pass through the body for some time. Similar conditions would occur if for instance, somebody is standing barefoot in a metal boat used as cathode, and touches the bare anode with his hand. In this case, the voltage would be between hand and foot.

Supposing a generator is installed in a wooden boat and one of the electrodes is connected with the metal casing of the generator. The fisherman touches this casing with one hand and catching the electrode with the other hand at the same time. He would be considerably safer if only one electrode is directly touched. The main drop of voltage occurs in the closest neighborhood of an electrode, underwater or in the earth so that a distance of only 0.5 m. from the electrode, the potential decreases to about 1/6 of that existing at the electrode itself. Therefore, if someone fell into the water from a metal boat used as cathode, with his feet 0.5 m. away from the live anode submerged under water and clinging with his hands to the metal boat, he would not suffer the total voltage of, for instance, 220 volts, but 220 V./6. His life would still be in danger but less so than if he made direct contact with both electrodes.

Supposing that both electrodes are under water with the voltage between them amounting to 300 V. If somebody in the boat dipped one of his hands into the water at a distance of 0.5 m from one of the electrodes and his second hand at a distance of 1 m from the same electrode, the potential difference between both hands in the water would amount to 15 V. which would be absolutely safe.

This distribution of voltage is independent of the conductivity of the water and to a large extent of the distance between the two electrodes.

Precautions

It is obvious that the greatest dangers occur when one or both electrodes are unprotected so that they can be touched with the hand or barefoot. In order to prevent accidents, only skilled electro-fishermen should be employed. Fishing should be directed by a responsible expert trained in a special course, who possesses authenticated proof of his qualification and permission from the responsible authorities to operate the gear. He should be assisted by at least one instructed person. People not concerned with the fishing should be kept away from the gear. In Germany, the electro-fishermen and his assistant have to know the VDE 0134 Anleitung zur ersten Hilfe bei Unfallen (Instructions for First Aid in case of accidents edited by the

Association of German Electro-Engineers) and must be familiar with at least one life restoring method. The fishing gear must be prepared carefully. The main electric cables especially have to thoroughly examined with regard to external damage. Above all the instruction of constructing electrical fishing gear and for its operation, are being prepared in Germany giving all necessary details.

Even the best preparations cannot prevent accidents which are caused by defectively constructed gear. Therefore the use of self-made equipment must be forbidden, and the construction of electrical fishing gear will have to comply with special instructions of the VDE. The operating handles, for instance, will have to be made of insulating material. The coating of the movable cables leading to the electrodes will have to be made of insulated material of especially high resistance against abrasion and buckling. An all pole safety switch interrupting the circuit in case of defective voltage or body contact will be prescribed. It will have to be installed as near as possible to the primary source of power to increase working reliability as well as safety and protection against accidental contacts.

Fuses will be provided in the line to cut out excessive current. In many cases a Totmann-switch at the catching electrode, which is handled by the electro-fishermen, is recommended. This guarantees that the electrode can not be switched on before all people concerned are ready for operation. In case of failures, the circuit can be interrupted immediately.

Chapter 12

Effect of DC and AC Shocks on the Carp Hatchlings and Young Adult

The behavior of *L. rohita, L. fimbriatus* and *C. mrigala* in their hatchling to early adult stages were classified as below.

Stage 0 (Normal)

The hatchling, irrespective of their species, on coming out of the egg shell, first settled at the bottom. They showed wriggling movement vertically upward to the water surface. On reaching the surface they dropped down to the bottom and lay on their sides till they shot up again. This type of movement continued till the yolk is completely absorbed and the alimentary system started functioning. The frequency of this type of vertical movement occurred at the rate of 12 to 16 times per minute. In 40 to 48 hours time the hatchlings started feeding. They did not lie quietly at the bottom, but started moving constantly by rapid active darting movement, characteristic of carp fry. At this stage they were seen to react to external stimuli to some extent and move away from the source of disturbance.

The horizontal swimming movement of young adults started with a side-wise lateral movement of the tail at the rate of 3-4 cm per second initially. The speed increased gradually with their growth and attained a speed of 6 to 8 cm per second, when their fins were fully developed within 15 days after hatching. The fright reflex of young fish to external stimuli was prominent only when they metamorphosed to young adults (20-25 mm). This reflex consisted of sharply turning away, moving in the tank in a coordinated swimming movement.

Reaction in DC Field

Stage I

The first reaction of the hatchling, irrespective of varieties, was initiated by their shooting up to the surface with a rotary movement along its axis, followed by increased vertical movement with the rise of current intensity (22 to 28 times per minute). They did not, however, respond to any other external stimuli at this stage.

Stage II

With the rise of field intensity irregular horizontal movements between the electrodes along with violent irritation were seen during this atage. The hatchlings, however, did not react to any other external stimuli. On switching off the current they dropped to the bottom.

Stage III

In this stage initially increased horizontal movements between the electrodes were observed which finally led to the floating of the hatchlings on the surface of the water with their heads up.

The time required by the hatchlings to congregate on the water surface varied with the field intensity and was 10 to 28 seconds for all the current densities and for all the species observed. Galvanonarcosis did not occur even though they were subjected to a current density of $1.445 \, \mu A/sq \, mm$ for 60 seconds at this stage. The response to external stimuli at this stage was similar to that of stage-II. On switching off the current flow, the hatchlings dispersed away from the water surface and settled at the bottom. The frequency of vertical movements after shock treatment, decreased to 3 to 5 times per minute.

Stage IV

With a further rise of field intensity, the hatchlings shot up to the water surface and were narcotized within 35 to 47 seconds and dropped to the bottom. This was observed as long as the current continued to flow. On cessation of current flow, they recovered and reverted to the vertical movements within 18 to 26 seconds at a frequency of 2 to 4 times per minute.

The pro-larva, 48 hours after hatching, with completely absorbed yolk and freedom to move and feed, exhibited the following behavior, when subjected to DC field, irrespective of their species

Stage I

They could perceive the underwater field with the jerks of the head and a change of position on switching on the current. At times they continued to move in a coordinated way during the passage of current. They were found to react to external stimuli at this stage by moving away from the source of stimulus.

Stage II

Violent irritation followed by irregular movements in the field were observed in the beginning in all the larvae which maneuvered to remain perpendicular to the field lines pointing their heads towards the margin of the field. In this position they gradually moved towards the negative electrode and clustered around it, within 8 to 15 seconds. At times, they floated and assembled, with their heads at the water surface near the negative electrode during current flow.Fish in this stage did not react to any other external stimuli. They could not be moved from their perpendicular position even though they were dispositioned with a glass rod. Normal coordinated movement of the organism was established immediately after the cessation of current flow.

Stage III

Being parallel to the field lines and heading towards the anode, the fish larvae moved violently towards the positive electrode, at a speed of 10 to 14 cm per second. Some of them showed forced movements towards the water surface and tried to remain perpendicular to field plane. At times,they also maneuvered to remain perpendicular to the field lines. Response to other external stimuli was similar to that of stage-II. On switching off the current, the fishes returned to their normal swimming condition immediately.

Stage IV

In this stage, the organisms were found to be relaxed and were not able to move of their own accord, lying on their side near the positive electrode. Accidentally a few (5 to7 per cent) heading towards the negative electrode could not maneuver to reach the anode in time, and were narcotized near the negative electrode. The effective period for narcosis varied from 10 to 65 seconds, depending on the field intensity. On stoppage of current flow they regained their movement within 2 to 270 seconds depending on the intensity of shock. After recovery, the fishes remained in a state of hypnotic trance with a slow voluntary movement, for 18 to 34 minutes before returning to their normal swimming movements.

The response and orientation of full grown young adults, 15 days old (20 to 25 mm), to the surrounding DC field with the increasing intensity were classified into the following groups for all the above mentioned species.

Stage I

Fishes lying parallel and heading towards the anode could feel the current exhibiting a fright reflex and a slow backward movement towards the middle of the field. Pointing their head towards the negative electrode, they showed irritation and jerking of the body, accompanied by increased movements of pectoral fins. Invariably they tried to orient themselves perpendicular to the field lines and stayed there till the current continued to flow. They also reacted to other external stimuli at this stage but at a lower intensity.

With the rise in current intensity, the pectoral fins started beating more rapidly without any movement from that position, that is, perpendicular to the current lines

Stage II

In this stage, the fishes moved violently towards the positive electrode at a speed of 13 to 17 cm per second, being parallel to current lines and heading towards the anode. Heading towards the negative electrode, however, they changed their position with body jerks at a right angle to the field lines and continued to stay in that position till the current continued to flow. They did not react to any other external stimuli and occupied their original position even though they were disturbed physically.

Stage III

On reaching the anode, the fishes incapable of any movement, lay on their side with their heads towards the positive electrode. The period required for such narcosis varied from 10 to 50 seconds, depending on the intensity of the current. Occasionally a few individuals, in their narcotic state, showed unbalanced movement and placed themselves perpendicular to current lines without any further movement as long as they were subjected to current exposure. On switching off the current the fishes regained their gill movements and pectoral beats and maneuvered to erect themselves in the swimming position within 5 to 45 seconds. Finally they swam away in a coordinated movement but at a lesser speed (5 to 7 cm per second).

Reaction in AC Field

Stage I

Embryos of *Labeo rohita, Labeo fimbriatus* and *Cirrhinus mrigala* immediately after hatching, could perceive the current. They exhibited semicircular movements in both horizontal and vertical planes, horizontal between the electrodes and vertical between the bottom and water surface on switching the current. With the rise of field intensity, the frequency of circular movement increased to 25 to 30 times a minute taking an elliptical path with a broader circumfirance. They did not react to any other external stimuli at this stage and settled down to the bottom with no further movement when the current flow was stopped.

Stage II

In this stage, the fishes moved violently between the electrodes, at a speed of 8 to 14 cm per second, parallel to current lines and settled at the bottom after 24 to 58 seconds. Response to external stimuli at this stage was similar to that of stage-I.

Stage III

Violent vertical movements of the hatchlings, accompanied by twisting of their body along its axis with their heads at the water surface, were noticed between the water column when they were parallel to the field lines. When perpendicular to the current lines, similar reaction of the hatchlings was noticed, but at a higher current intensities. Response to any other external stimuli was absent.

Stage IV

Followed by the reaction of stage-III, the hatchlings dropped to the bottom, incapable of any movement and continued to remain in this quiescent state during the passage of current. This type of narcosis set in within 2 to 20 seconds, depending on the current intensity. On switching off the current, they recovered to their normal swimming condition within 12 to 105 seconds.

Stage V

With further rise in current intensity, the hatchlings underwent violent contraction of body within 2 to 10 seconds and were incapable of any movement. During this condition, their bodies remained in a contracted state, accompanied by the inward bending of the notochord. Recovery from this stage was noticed within 10 to 25 seconds after the stoppage of current flow which at times was followed by sporadic swimming and settling at the bottom. The notochord remained bent in some cases even after they recovered from this stage.

Stage VI

Followed by the tetanus of the body, 85 to 100 per cent of the hatchlings died with the bend notochord and opaque yolk (Figure 12.1) within 8 to 20 seconds depending upon the intensity of current exposure.

The response of 48 hour old larvae in the increasing AC field have been grouped as below for all the species tested;

Stage I

Organisms felt the current with a jerk of the body and a change of position (from parallel to perpendicular to the current lines during the flow of current. After switching off the current, they showed normal coordinated movement. In this stage they were found to react to other external stimuli and moved away from the source of disturbance.

Stage II

They showed erratic movement for 5 to 10 seconds during the passage of current and then assembled at the center of the field floating near the water surface with their body axis perpendicular to the field lines and their heads at the water surfaces. Response to other external stimuli at this stage was minimum. In this stage they were not narcotized even though the current flow continued for 60 seconds. On switching off the current, they dispersed in the field evenly and settled at the bottom.

Stage III

Violent sporadic movements between the field for 3 to 5 seconds followed by cessation of voluntary movements of the larvae took place in this stage, resulting in their sinking to the bottom throughout the field within 15 to 25 seconds. Organisms that lay parallel to the field lines showed no movement at all, while those perpendicular to field lines showed vibrations of the tail. After discontinuing the current flow, they started moving slowly in the beginning in an unbalanced manner. Some of them exhibited a rotary movement of the body with their heads at the center. Recovery of normal movements was completed within 30 to 300 seconds.

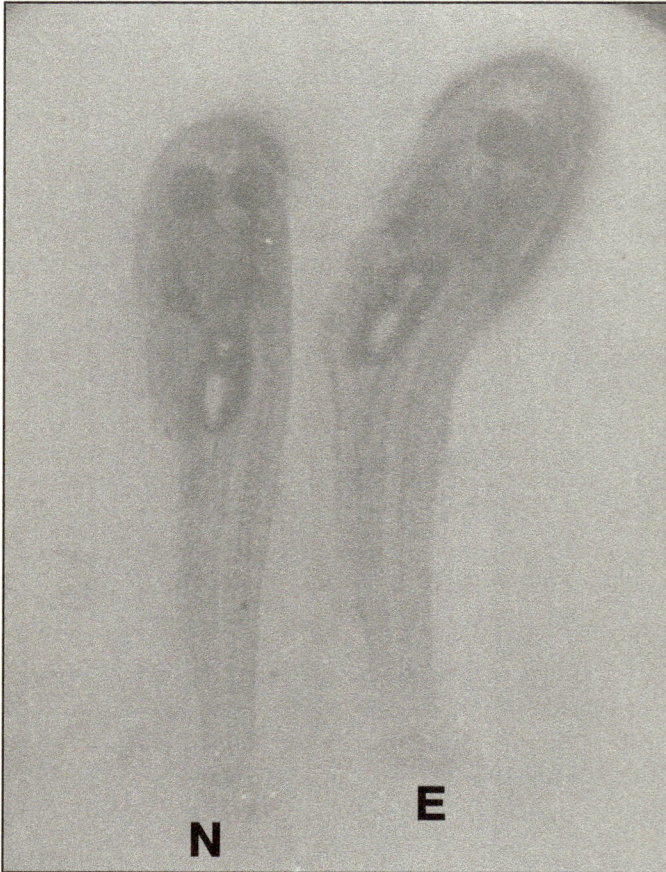

Figure 12.1: Hatchlings with bend notochord and opaque yolk (immediately after shock treatment, 10.5 x). N: Normal; E: Treated.

Stage IV

Within 4 to 9 seconds of current flow the organisms showed cramping of their bodies in their original position without able to move and settled at the bottom. Switching off the current immediately after tetanus, enabled the organisms to recover from this state within 85 to 97 seconds.

Stage V

80 to 95 per cent of the organisms in the titanic state, if allowed to continue to be subjected to the current for 15 to 27 seconds, died with curvature of the body.

Adult fishes of 20 to 25 mm, when treated in the AC field showed the following behavior under broad classification irrespective of species.

Stage I

Fishes could sense the surrounding electrical field with jerks of body and a change of position. Their gill movement increased by 34 to 46 per cent. At times, some individuals jumped out of the field. They were found to move away from the source of disturbances when subjected to other external stimuli at this stage. On switching off the current, they returned to their normal swimming condition immediately.

Stage II

With the current rise fishes moved violently between the electrodes, when they were parallel to the current lines, and tried to place their body axis perpendicular to the field lines without any further movement till the flow of current continued. Being perpendicular to the lines of force, they remained in the same position with the increasing field strength and did not respond to any other external stimuli. Their activities slowed down after the current shock.

Stage III

In a higher field intensity, fishes underwent narcosis, followed by violent movement between the electrodes within 16 to 20 seconds and settled at the bottom by closing the gill cover. Gill movements reappeared within 10 to 58 seconds and they recovered to their normal swimming position within 87 to 122 seconds after cessation of current flow. The gill movement after recovery, increased by 30 to 32 per cent of the normal beat.

Stage IV

At a still higher current density, their bodies underwent violent contraction and could not move from their original position. They sank to the bottom with expanded fins, gill covers and bending of the body within 7 to 14 seconds of current flow, and did not exhibit any type of movement. Gill movements started, 6 to 15 seconds after stoppage of current and they could maintain their equilibrium within 104 to 125 seconds. Bending of vertebral column was noticed in 30 to 35 per cent of the cases and 2 to 5 per cent decrease in gill movement was observed in organisms after recovery from the tetanus.

Studies in the three different stages of their growth, (newly hatched out embryos with yolk, young larvae with fully absorbed yolk and young adults) to observe the electrical responses at different stages of the development of their nervous and locomotor systems revealed that the response of pro-larva to early adult at three stages of development and their orientation in the DC field with the increasing field intensity, indicate that the perception of the electric field at their early age (hatchlings, with undeveloped sensory and locomotor organs) were characterized by contortion of their bodies, abrupt and disorganized swimming movement and cessation of voluntary movements. They did not exhibit a distinctively directional galvanotropic response at this stage (Biswas, 1977).

A similar response was observed by Godfrey (1957), when salmon alevins contorted violently and swam sporadically without orienting in relation to the direction of the source of current, when they were given DC shocks of 575 volts at 1.5 amperes.

According to Halsband (1968) the galvanotropic response of a directional character is initiated by the reflex center, located at the mesencephalon-cerebellum level and also by the medullary sensory path. The newly hatched larva with undeveloped encephalon and afferent paths failed to exhibit directional galvanotropic response.

The advanced larvae, besides increased activity congregated near the negative electrode and placed their body axis perpendicular tp the current lines with the increasing field intensity. They also exhibited anodic narcosis in higher current intensity, a step advanced in galvanotropic behavior, which was due to developed electro-sensory organs as the growth continues. The concentration near the negative electrode cannot be called as true cathodic galvanotaxis since the larvae did nor swim fast with short undulations towards the cathode, but drifted near it in a perpendicular position, pointing their head towards the margin of the field.

The young 15 day old adults of 20 to 25 mm length responded with retarded swimming during fright reflex, forced swimming towards the positive electrode, and galvanonarcosis with the increasing field strength, characteristics of a fully grown fish, confirmed by Halsband *et. al* (1960). Irrespective of species, *L. rohita, L. fimbriatus* and *C. mrigala* have been found to react for directional galvanotropic responses when they attained a length of 20 to 25 mm, fifteen days after hatching.

The threshold current densities for the different reactions of fishes at three different size and age did not show any significant difference even though a difference of 15 to 20 mm in length existed between them, which was probably due to undeveloped sensory and locomotor organs in the first two groups for which they failed to show true taxis and narcosis in time.

Scheminzky *et. al* (1941) observed the initiation of the first reaction, galvanotaxis and galvanonarcosis of *Phoxinus laevis* with the rise of current densities at the ratio of 1 : 16 : 30 : for fishes of 36 mm length. The ratio of current densities for the responses of newly hatched larvae at successive stages was 1:1.84:2.49:3.13 for *Labeo rohita*; 1:2:2.59:3.28 for *Labeo fimbriatus* and 1:1.9:2.39:3.16 for *Cirrhinus mrigala*. These ratios altered to 1:2.44:4.16:6.32; 1:2.74:4.41:5.36 and 1:2.74:4.41:5.36 for the above species respectively as the hatchlings grew older and finally the ratio of reaction threshold for the young adults was 1:2.41:3.14 for *L. rohita*; 1:2.63:3.21 for *L. fimbriatus* and 1:2.95:3.81 for *C. mrigala*, indicating the requirement of broad spectrum threshold values for the fish larvae with rudimentary sensory and locomotor organs than the earlier and later stages of development. This may be explained as to the delicate condition of the hatchlings and the perception of higher potential between the head and tail of the young adult.

The effective period for responding to electrotaxis in any form varied inversely with the progressive development of the embryos which is probably due to increase in length with development. Meyer-Waarden (loc.cit.) stated that since large fishes receive a greater voltage in water than smaller ones, they could be influenced more quickly. But there was no significant correlation with the effective period for narcosis and the period of recovery with the development of the organisms.

When these species in identical developmental stages were subjected to uniform AC field in which the organism successively faces the anode and the cathode 50 times per second, the new born hatchlings exhibited semicircular vertical and horizontal swimming movements, narcosis, tetanus and finally death with spinal curvature with the rise of field intensity. Similar explanation as in the case of DC field can be put forth for these disorganized response by the new born embryos, where their sensory and locomotor organs were not developed and consequently their reaction was almost one and the same for DC and AC field. The only exception of titanic condition and death with spinal curvature at higher intensities was due to the severity of AC shocks over that of DC.

The appearance of transverse oscillotaxis, as described by Halsbant *et. al* (1960), Lamarque (1963) and Meyer-Waarden (1955) in 48 hour old fish larvae, when they took a transverse position in the AC field, indicated further improvement of electro-sensing and locomotor function of the organisms with the advancement of age.

The early adults of all the three species exhibited reactions, characteristic of AC shocks, namely, fright reflex as explained by Bodrova and Kraiukhin (1960), longitudinal oscillotaxis as reported by Lamarque (1963), transverse oscillotaxis, slackening of the body muscles without voluntary movements and finally tetanus with the rise of field intensity. This goes to support that organisms of 20 to 25 mm can respond to AC field in a systematic way.

In DC field, the reactions in different stages were initiated with the progressive rise of field intensity irrespective of species and their developmental stages. The ratio of threshold current densities for the first reaction,oscillotaxis, narcosis and tetanus was 1:2.71:3.39:5.05 for *L. rohita*, 1:2.72:3.40:4.90 for *L. fimbriatus* and 1:2.78:3.50:5.02 for *C. mrigala* in the case of new born hatchlings. These ratios changed to 1:3.02:5.74:9.78 for *L. rohita*, 1:3.19:6.15:10.63 for *L. fimbriatus* and 1:4.75:8.99:15.13 for *C. mrigala* when the larvae were 48 hours old after hatching. The threshold current density for the adult fishes for different reactions indicated that oscillotaxis, narcosis and tetanus of *L. rohita* averages 2.5, 4.4 and 9.48 times that for the first reaction. The corresponding values were 2.61, 4.23 and 8.5 times for *L. fimbriatus* and 2.61, 4.28 and 8.23 times for *C. mrigala* respectively. The requirement of the lowest reaction threshold in the adult stage, irrespective of species, was due to their increased length in addition to well developed electro-sensing and locomotor function, when the organism was subjected to higher potential difference between the body extremities and consequently influenced earlier at lower reaction thresholds. In both DC and AC field, the larvae exhibited the reactions in threshold current densities of wider range as compared to their requirement in hatchlings and young adults.

AC shocks were proved to be very severe to hatchlings, when they underwent narcosis and tetanus within 2 to 20 seconds leading to 85 to 100 per cent mortality within 8 to 20 seconds in titanic condition.

Forty eight hour old larvae were not only more resistant to AC shocks, but also developed better electro-sensing, confirmed by the fact that they responded for oscillotaxis within 5 to 10 seconds and took a longer period for narcosis, tetanus and death of 80 to 95 per cent organisms.

The young adults were still more resistant in AC field and did not die even though they had undergone narcosis, tetanus and spinal curvature in 30 to 35 per cent of the cases (Biswas, 1977).

Lamarque (1963) stated that alternating current provoke violent tetanus which causes nervous exhaustion. Depending on the exposure time to the alternating current remnant paralysis may follow the titanic phase. He further stated that, this paralysis, which results from synaptic exhaustion, should be avoided for survival and recovery of fish under experiment.

The rate of reaction thresholds, for *L. rohita* larvae in AC field varied inversely with the rise in conductivity of the media. According to Holzer, cited by Denzer (1956), the potential difference between the head and tail is an important factor for determining the reaction of fish, which varied with the relationship between resistivity of fish body and that of the conducting media. In a higher conductive media, a little amount of current intensity is impressed on the fish body and consequently a higher value of current density is required to influence it.

Effect of DC, AC and Impulse Currents upon the Metabolism of Fish

Halsband (1955) had studied the influence of direct, alternate and impulse currents on the fish metabolism. He worked on trout (*Trutta iridea*, and the intensity of metabolism was determined by measuring the consumption of oxygen, the breathing frequency and the intestine temperature.

After the electrical stimulation, all kinds of current developed a changed intensity of metabolism with the experimental fishes. Small densities of current (limits for the stages of excitation and electrotaxis) caused an increased oxygen consumption, breathing frequency and intestine temperature, whereas greater densities of current (limits for the electronarcosis) achieved a retarded metabolism.

The normal values of the intensity of metabolism are restored after a treatment with direct current after 70 minutes on the average, after a treatment with alternate current after 120 minutes on an average, and when stimulated with impulse current after 20 minutes on the average.

Alternate current has the greatest and longest percentage share in the influence on the metabolism, whereas impulse current has the smallest effect.

The exciting effect of the electric current to the intensity of metabolism does not only change into a paralyzing effect with the increasing density of current, but it occurs also when the period of the electrical treatment is prolonged.

The change of the intensity of metabolism after the electrical treatment can be caused by several factors. Direct effect of the current to the regulative mechanism of the breathing activity, influence upon the required muscular activity and motorial movements of the blood vessel system, changes in the chemical concentration of the tissue liquid and the cellular substance.

Effect of DC and AC on the Blood Cells of Fishes

The erythrocytes of carp (*Cyprinus carpio*) and trout (*Salmo gairdnerii*) were first of all examined under the microscope in a direct-current field, various current densities and rates of flow being applied. In the first reaction stage a migration of the blood cells to the anode occurs (cataphoresis), in the second stage the cells change in shape from oval (Figure 12.2) to round (Figure 12.3) and in the final stage the cells disintegrate (Figure 12.4).

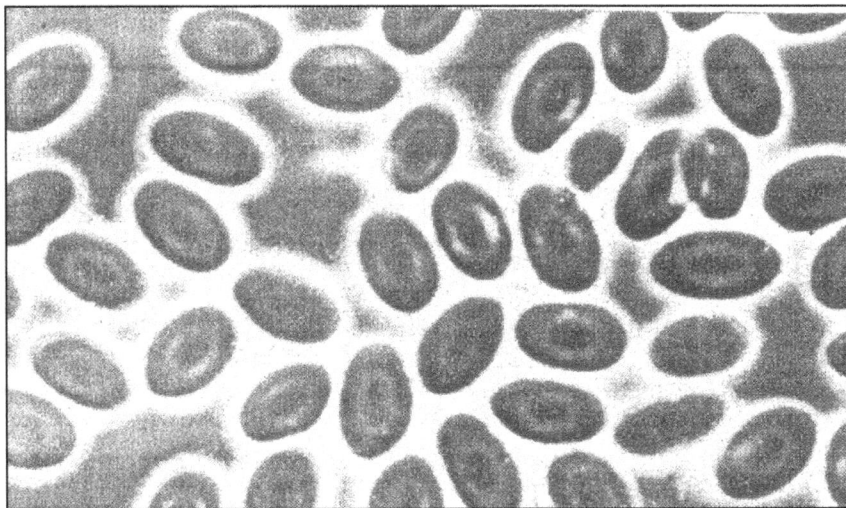

Figure 12.2: RBC (Erythrocytes) of trout (Normal form).

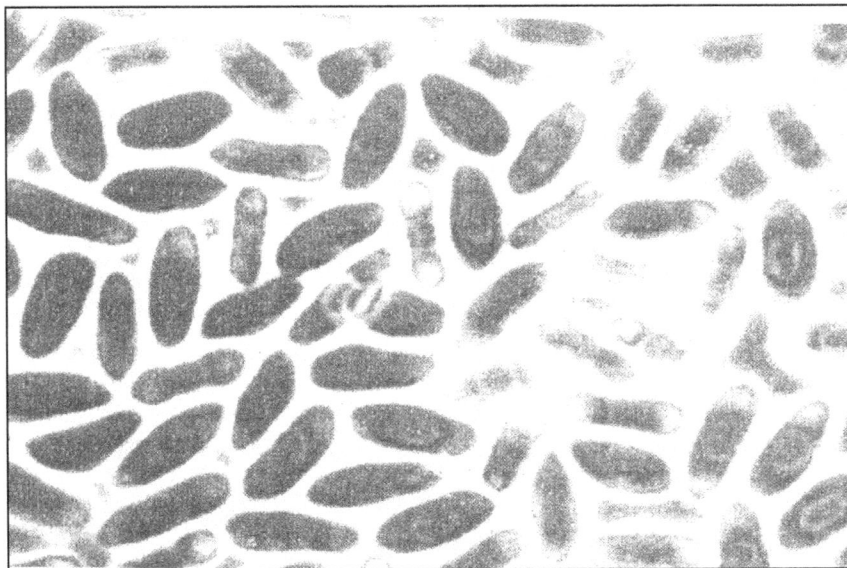

Figure 12.3: Change in shape of erythrocytes at the second stage of DC field.

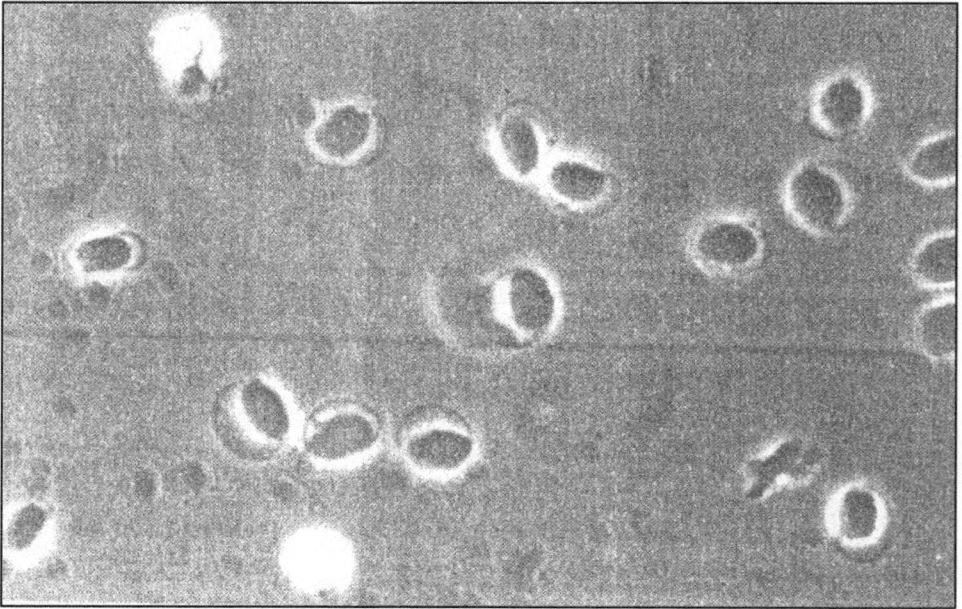

Figure 12.4: In the third stage, the blood cells of trout became small prior to disintegration.

When equal current densities are used the blood cells of trout undergo these changes at lower rates of flow in comparison with those of carp.

In order to examine the dependence of the reaction intensity on the degree of hydration in the blood cells, the experimental fish were introduced in various saline solutions.

When solutions raising the degree of hydration in the cell tissue were employed the alterations in the electric field set in more quickly. The resistance of the cells to the electric current was thus lower.

When solutions lowering the water content of the cells were employed an increase in the resistance of the blood cells was observed.

Moreover, examination was made of the effect on the fish of several hours of permeation with direct, alternating and pulsed current. The influence of x-rays on the number and surface of the red blood corpuscles of trout was also investigated.

Application for 18 hours of direct and alternating current with a density which just sets off the first reaction (twitching of the fish) induces changes in the number and surface area of the erythrocytes. However, several hours after permeation the erythrocytes return to normal. After 18 hours permeation with pulsed current, the number and area remain at their normal level.

Irradiation with roentgen units of 750 to 2500 r causes changes or damage to the blood cells which can not be corrected subsequently.

The necessary densities of current effecting a change in the blood cells are the multiple (about 1000 to 2000 times) of those values necessary for attracting or stunning the fish in the practical electrical fishery. The densities of current used in electrical fishery thus do not cause any direct changes of or damages to the blood cells of fish.

Chapter 13

Choice of Current Type

The choice of best current for electric fishing is based on both physiological and practical reasons.

From the physiological point of view, the current to choose will be that which does not fatigue the fish, cause minimal damage, attracts it to the electrode and saves energy consumption. For physiological effects the currents can be classified, from worst to best in the following order :

   (1) Alternating current, 50 or 60 Hz,

   (2) Alternating current 300 Hz,

   (3) Condenser discharges,

   (4) Half-wave rectified AC, 50-60 Hz, single phase,

   (5) Full-wave rectified AC, 50-60 Hz, single phase,

   (6) Square-wave pulses, pulse duration > 1 ms, frequency < 200 Hz,

   (7) Quarter-sine waves, pulse duration 5 ms, frequenct 50 Hz,

   (8) Quarter-sine waves, pulse duration 5 ms, frequency 100 Hz,

   (9) Square-wave pulses, pulse duration > 50 ms, frequency < 10 Hz,

  (10) Square-wave pulses, pulse duration > 0.25 ms, frequency > 400 Hz,

  (11) Half-wave smoothed rectified AC, 50-60 Hz, single phase,

  (12) Full-wave smoothed rectified AC, 50-60 Hz, single phase,

  (13) Half-wave rectified AC, 300 Hz, single phase,

  (14) Full-wave rectified AC, 600 Hz, single phase,

  (15) Half-wave rectified AC, three phase,

  (16) Full-wave rectified AC, three phase,

  (17) Pure DC

Examples of practical reasons include;

☆ The combination of high conductivity water and back pack gear imply the use of pulsed current which can save energy and also reduce the weight of machine;

☆ Low conductivity water conduces the same conclusion, as in this situation it is necessary to use a very high voltage,

☆ Difficult local fishing conditions make it necessary to attract the fish to the electrode, that is, DC and its family of rectified currents,

☆ A range of fishing conditions which requires simultaneous use of several current types

All these currents have been used by Lamarque, (1990) in the field for a range of species and field conditions. The reasons behind the sequence are given below.

To produce a good electrotaxis it is necessary to avoid tetanus. To avoid tetanus it is necessary to stimulate the nerve cells and not the fibers as well as to protect the motor nerves against hyper-reflexivity. From this point of view DC is the best current, but, the energy consumption is high and threshold reactions are induced at greater voltage than variable currents. Moreover, DC can not be transformed. It appears therefore necessary to use pulsed current derived from AC which can be transformed.

Long duration pulses, with a low frequency (4 to 10 Hz) can produce summations of the body cells (# 4) thus acting like DC but reducing the over-stimulation produced by the hyper-reflexivity (# 2). However, this leaves the possibility of direct stimulation of the nerve fibers. With a pulse duration of about the useful time (1-3 ms), at any frequency, the worst stimulation effects will occur as the nerve fibers are fully excited without any anodic protection. A strong tetanus will occur.

With a pulse duration somewhat shorter than the useful time (< 0.25 ms for the voltage used in electric fishing) direct stimulation of the nerve fibers does not occur. If the frequency is very high in comparison to the organisms natural frequencies, between 400 and 1000 Hz, the DC component of the short pulses can produce summations on the body cells and a DC mechanism will occur. Such a current was verified in practice by Lamarque (1976a) but has never been used by other scientists except Burnet (1959), in the laboratory. He found that the least tetanizing pulsed currents had frequencies between 1000 and 2000 Hz and pulse durations between 0.2 and 0.1 ms. Unfortunately he did not use the current in practice because of the technical problems at the time.

To get the best summation effect it is necessary to use currents with strong DC component, that is, square pulses. At a frequency of 1000 Hz, 10 per cent duty cycle and a pulse duration of 0.1 ms the behavior is that of DC but the threshold reaction is higher than DC. At 400 Hz and 10 per cent duty cycle with a pulse duration of 0.25 ms the reaction threshold is half that of DC (Lamarque, 1976a), showing that some effect of the pulse direct stimulation still exists. Nevertheless, with such a current, anodic electrotaxis is good and the fish are not fatigued. A better frequency could be 500-600 Hz however, this current has not yet been tested. Such a current represents a compromise of effects between DC and pulsed DC and saves 90 per cent of the energy.

Rectified AC can be considered as pulsed DC and DC according to the frequency and the undulation rate. It is possible to transform rectified AC and consequently increase the working range of the machine.

To assess the energy consumption of different currents, the following fishing conditions should be considered;

☆ Water conductivity 1000 micro siemens/cm;

☆ Anode ring 40 cm diameter;

☆ Large cathode and

☆ Voltage 300 V

In this conditions the resistance ® between the electrodes will be about 10 ohms. The energy consumption for DC will be;

P = V square/R = 90000/10 = 9 kW

With a long pulse duration of 50 ms, low frequency at 10 Hz and a duty cycle (dc) of 0.5;

P = V square/R x dc = 90000/10 x 0.5 = 4.6 kW

With short duration pulses (0.25 ms) at high frequency (500 Hz) and a duty cycle of 0.1;

P = V square/R x 0.1 = 90000/10 x 0.1 = 0.9 kW;

Thus, this latter current saves 90 per cent of the needed power before consideration is given to threshold differences.

Chapter 14

Reactions and Behaviors of Fish in Heterogeneous Field

Visible signs of fish behavior and their physical reactions subjected to an electric field was observed with a definite sequence of reactions with an increase of field strength. Investigations were carried out with different types of current, various species and length of fish, their physiological condition and environmemt parameters (temperature and water conductivity). All most all these tests were held under laboratory conditions (in aquaria) and what is of particular significance here in homogeneous electric fields. The experiments aimed at the determining voltage gradients, current densities or head-to-tail voltages necessary to produce the first reaction, electrotaxis or electronarcosis. For getting uniform results, it was necessary to place fish exactly along the lines of current flow, that is, perpendicularly to the equipotential lines. An electric field affects fish more strongly when the head-to-tail voltage is higher. If a fish finds itself just along the equipotential line, that voltage equals 0, and the fish hardly feels any electrical stimulus, even if the electric field is strong.

Homogeneous and Heterogeneous Fields

An electric field is homogeneous when the current densities at all its points (size and phase-angle of the vector) are the same. The field pattern consists of equally-speced parallel equipotential planes (Figures 14.1 and 14.2).

Any electric field that is not homogeneous may be called heterogeneous. Figure 14.3 shows such a field made in an electrolytic tank by means of two vertical bar electrodes A and B. Lines that join the electrodes indicate the direction of current flow. Such lines always cross the equipotential lines at right angles, so that it would

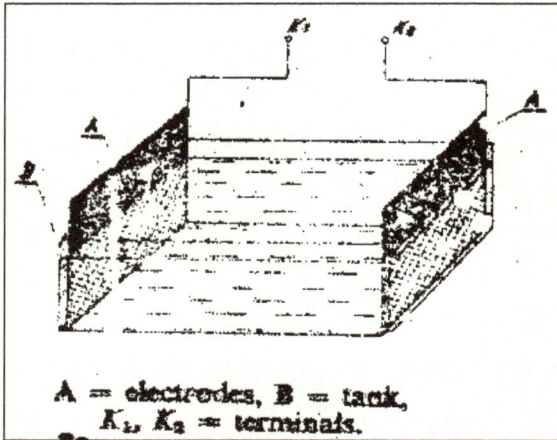

Figure 14.1: Electrolytic tank made of insulating material with two plane electrodes used to produce homogeneous field.

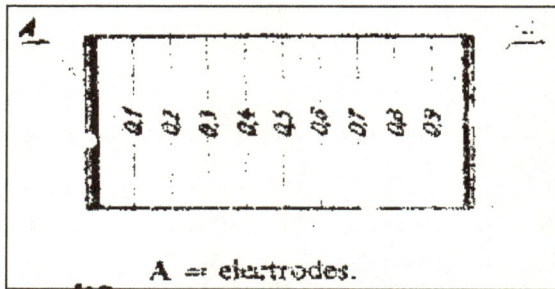

Figure 14.2: Lines of equipotential of the homogeneous field in the tank shown in Figure 14.1 (Figure on each line gives its potential in relation to the voltage between electrodes).

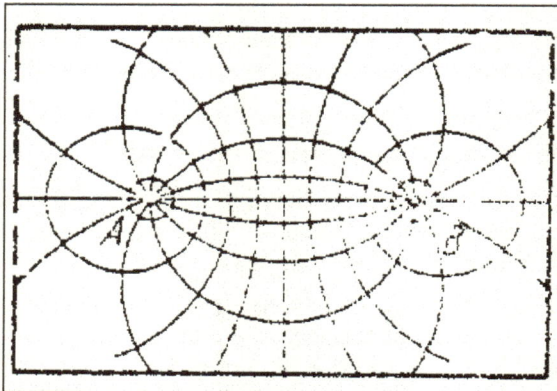

Figure 14.3: An example of heterogeneous field.

be possible to visualize the direction of current flow having only the lines of equipotential.

Electric Fields with Single and Multi-Phase Current

Electric fields containing fish screens used practically are adequately illustrated by the equipotential line patterns. Such lines, independently from the kind of electric current (direct, alternating, pulsating currents etc.) hold in current duration the same place in space. In such a sense they are stationary fields.

Chmielewski (1964) has proposed a new type of electric fish screen with a single-row electrode array supplied from 3-phase mains through a transformer giving multi-phase current. The electrodes are wired according to a special wiring pattern which has been patented.

The electric field of the new screen, has special characteristics. It has no stationary equipotential lines. The field can be interpreted as having equipotential lines that move with a given velocity along the row of electrodes in such a manner that it can be said the field is moving.

Space Properties of an Electric Field

It has been stated (Newman, 1959) that the electric fish screen caused a fish response in the nature of a simple avoidance of an unpleasant stimulus rather than a forced reaction to the directional properties of the field. Up to that time it was assumed that the field polarity, that is, the galvanotropic effect (galvanotaxis) was indispensable for efficient stoppage of fish.

Efficiency of an electric fish screen depends on the characteristics of an electric field in water. This can be specified as follows:

1. Current time characteristics;
2. Field properties in space

All specialists agree that a pulsating current ensures better efficiency than steady direct or alternating currents. The pulsation of an electric current is of main importance, but it is supposed that the shape of a pulse (unidirectional, bidirectional, square-wave, half or quarter sine-wave, exponential, pulse including some full periods of sine-wave etc.) only slightly affects the final efficiency of a screen, provided the pulse and pause durations are equal. At the same time great importance is attached to the space characteristics of an electric field. These special characteristics have been underestimated in almost all studies.

Thus it can be concluded that;

1. Interruption of electric current enlarges the orientation zone of fish.
2. Larger fish feel electrical stimulus farther away from the screen, for such fish orientation zone is bigger.
3. Slow increase of voltage gradient make the orientation zone large.

The efficiency of an electric fish screen becomes greater, the slower the voltage gradient rises, that is, the wider the zone of orientation in which a fish maintains its orientation ability. In such circumstances the fish has more time and space for realizing its defensive reaction and finding a place for safety. If the width of orientation zone is insufficient then, soon after the first contact with an electric field, the fish can immediately penetrate into such a strong field from which it will be unable to escape.

The differences in space characteristics between single and multi-phase current fields are of great importance to fish placed in it. In a stationary field fish can always find a position (along a line of equipotential) in which body voltage drop is nil. In a moving field such position does not exist because of the absence of stationary equipotential lines. In every position the fish is exposed to a certain head-to-tail voltage.

In any heterogeneous field, of course, fish may avoid high voltages by rapidly swimming away from the electrode array.

Fish Behavior around Heterogeneous Field

To apply a heterogeneous electric field successfully in an electric screen in a given situation, it is first necessary to have some understanding of the behavior of the fish species which is to be managed. Fish move for a variety of basic reasons; foraging for food, migration, reproduction, escape from predators. The drive stimulating these actions are so strong that fish responding to them are not deterred by extreme hazards. Salmon attempting to leap rapids to reach spawning grounds are a good example.

Fish encountering heterogeneous electric field produced by electric screens, under such circumstances, will make repeated attempts to pass and sooner or later will either succeed or be harmed in the attempt. Electric screen would appear to be unsuitable for such situations and should perhaps be used to modify rather than prevent a particular form of fish behavior.

For example, fish may be sucked into turbine inlets involuntarily, not perceiving that by drifting with the current they are in danger. If there is an alternative escape channel, then it may be possible to provide a stimulus in the turbine inlet which will deter the fish from entering. In cases, where fish are not migrating, but are moving randomly within an area in search of food, it may be simpler to confine them successfully by an electric screen, especially in the absence of water current.

The exploratory behavior of several species of fish near screens has been reported by Stewart (1981) and Wardle (1986). Observations were made on fish encouraged to approach screens by responding to a feeding stimulus and on fish approaching and exploring screens without inducement. The main finding was that the visual appearance of the barrier was critical to its effect. It was found necessary to mark physically the position of an electric screen in order to make it effective. Rope was used to mark the screens. Fish approached any screen with care and explored it for varying periods, from minutes to hours depending on the species. If the screen was harmless to the fish, they would then pass through and insubsequent encounters, approach and pass through rapidly.

Electrified screens, which could harm the fish, remained effective and eventually the fish ceased to approach. With a visually unmarked electrified screen, fish swam into the field, received a strong muscular stimulation and escaped by swimming vigorously, either back or through the field. Such a screen was inefficient, apparently because the fish were unable to identify and avoid the electrified zone.

The basic reactions of fish to electric fields are well known and may be described, for increasing field intensity, as; perception, muscle stimulation and flight, electrotaxis and electronarcosis. Most electric screens rely on relatively low intensity fields, providing enough stimulation to influence behavior but avoiding harm to the fish. Strong fields, may however, be encountered near the electrodes which can be harmful. The field distribution can be highly non-uniform and fish entering it appear to be unable to discern the direction of increasing field intensity. They may, when stimulated, swim into rather than out of the electrified zone. Screens with slowly increasing field gradients have been tested to try to overcome this limitation, but with mixed success (Hadderingh and Jansen, 1990).

Screens have been tested on sharks, round fish, flat fish and shell fish. Given the wide range of habitats in which these species are found, it is likely that there will be many strategies for using electric fields to confine or divert fish. Since the electric field is only one of many stimuli which a fish will experience, the reaction to a weak field, less than the threshold for electrotaxis, will be conditioned by the total environment. The factors which are likely to be most important are the water speed, the water temperature and the visibility of the screen. Ideally, fish behavior in the chosen area should be studied before any decisions are taken about the detailed structure of the screen.

Fish Reactions in the Electric Field of Multiphase Current

Having observed the fish reactions, as the voltage gradients were increased, it has been concluded that there might be three basic phases of reaction of fish exposed to the electric field of multiphase current.

1. The first reaction- the fish quivers with its whole body at the moment the current is switched on.
2. Continuous motion- during the whole period of current flow, the fish swim restlessly and, what is significant, without interruption all over the field.
3. Electric shock- at the beginning the exposed fish moves very quickly and spasmatically; then at the end of the test it turns on its side and becomes immobile.

The continuous motion, appearing at voltage gradients close to those for the first reaction is very characteristic for a multiphase current field. The fish, at the very impulse,, independent from the position of its body, is exposed to a certain voltage. This is caused by an absence of stationary equipotential lines. Consequently every pulse initiates a new motion.

Such reactions does not occur in electric fields that have stationary equipotential lines. The necessary tests were carried out in the field of multiphase current, fed also by alternating current (50Hz), which was interrupted with a small frequency equal to 3 Hz, and a pulse duration of 0.11 second. It was stated that at the moments when the fish was perpendicular to the lines of equipotential, its behavior is the same as in M6 field. However, the exposed fish takes all other position too. In those cases the electric field effect is not so strong. It had even happened that the fish, when oriented along the equipotential line, stopped its motion for a few seconds, or even for the whole exposure time. Here then are pulses that do not stimulate any movement. If the strength of the electric field is increased then the continuous motion could again be produced, but it would happen at voltage gradients close to the electric shock, that is, the fish would be overcome and not stimulated to escape from the field.

According to opinions expressed earlier, the reduction of zone width between the first reaction and electric shock could result in making the orientation of fish more difficult.

Chapter 15

Role of Olfaction in Fish Behavior

Like all vertebrates, fish have two well-defined sensing systems, taste (gestation) and smell (olfaction). Evolution has favored the development of these systems for several reasons. First, chemical cues are ubiquitous in the aquatic environment. Organic compounds are released in substantial quantities by aquatic organisms as a consequence of osmoregulation, metabolism and eventual death and degradation. Second, water is an "universal" solvent, allowing the vast majority of these chemicals to dissolve, making them available to function as biologically relevant cues. Lastly, because of the appearance of highly specific protein receptor sustems very early in evolution, extremely complex mixtures of chemicals (both odorants and tastants) can be distinguished, giving these cues an extraordinarily high " information content".

The cellular and biochemical bases of smell are highly conserved among vertebrates, and fish have traditionally served as important models for understanding the basic physiology of the system because of the ease with which they are studied and the fact that, at least for olfaction, the range of cues detected is probably less than for tetrapods. Although tetrapods have two olfactory systems (the main and accessory systems), fish possess just one which contains both of the receptor cell types found in the mammalian systems. The ranges of chemical stimuli detected by fish olfactory system are quite different from those detected by their mammalian counterparts (likely a reflection of the cues found in the aerial and aquatic environment), and more restricted. It has been estimated that the fish olfactory system can distinguish about hundred individual chemical stimulants, an order of magnitude less than that hypothesized for the mammal olfactory system. Those species of fish which have served as important models of chemosensory function are, catfish, goldfish, and salmon- all freshwater teleosts. Unfortunately, little is known about the chemosensory abilities of the other 24000 or more species of fish, some of which most likely have different chemical sensitivities and specificities because of the varied nature of their behavioral ecology and taxonomy.

The olfactory system mediates responsiveness to a much broader range of stimuli including learned food odors, geographic locations (learned and innate orientational cues), and conspecifics (pheromones). The variety of stimuli detected by the olfactory sense thus appears much greater than that of the taste sensing system, although both systems are sensitive to amino acids and nucleotides. The olfactory system of fish is fundamentally similar to that of other vertebrates, at least at the cellular level. Unlike in taste, olfactory sensivity is mediated by a single cranial nerve (the olfactory nerve) whose neurons are unique both because they are primary sensory neurons which project directly to the brain, and because they regenerate throughout the life of the animal.

The Olfactory Organ

In fish, the olfactory receptor cells are located in the olfactory epithelium which covers much of the surface of the olfactory rosette, a structure found within the olfactory chambers on the fish's rostrum. Although the size and shape of the rosette varies greatly across different species, in most instances it has a longitudinal ridge (raphe) with two rows of olfactory lamellae radiating from it, thus increasing its surface area enormously. Significant variations from this pattern include circular and semicircular rosettes, and rosettes without lamellae. Most fishes have paired olfactory chambers, the exception being the cyclostomes which are monorhinic (having only one olfactory organ and associated nostril). Generally the olfactory chamber has both incurrent and excurrent nostril. Often a ventilation chamber is also associated with this structure, and it can be pumped to evoke "sniffing" The shape and size of olfactory organ, the size and morphology of the nostrils, the position of the olfactory organs with respect to each other, and their spacing on the head vary great among species, presumably reflecting the variety of ecological conditions under which fish live. Examples of variations in this structure are found in hammerhead shark whose olfactory organs are located in rather superficial pits at each end of its T-shaped rostrum, the flying fish whose olfactory epithelia are located on an external protrusion, and the freshwater eel whose enormous olfactory organs are located in large sacs, the ventilation to which are controlled by well developed nasal tubes. Sexual dimorphism is also evident among the Myctophiformes and Lophiformes, the males of which have much larger olfactory organs, presumably reflecting the importance of pheromones in the deep sea.

Olfactory Stimulants

The understanding of the nature of odors recognized by the fish olfactory system is largely derived from behavioral and electrophysiological studies in conjunction with biochemical analysis. The compounds that function as olfactory stimulants (odorants) in fish tend to be small metabolic byproducts which are both water soluble and which if detected, have some inherent biological meaning (for instance, amino acids signal a food source). Four major classes of odorants have now been clearly described in several species of fish; (1) amino acids; (2) bile acids; (3) gonadal steroids and derivatives; and (4) prostaglandins. Various alcohols, amines, carboxylic acids, nucleotides, and aromatic hydrocarbons have also been found to have olfactory activity in various fishes.

Amino Acids

Amino acids are important olfactory stimuli in fishes. All fish tested have been found to smell L-amino acids. L-amino acids serve as a cue for food recognition, amino acids have also been suggested to serve as social cues. All fish tested have been found to detect the majority of natural L-amino acids. The olfactory systems of common carp and several salmonid species can discriminate between multiple classes of amino acids.

Bile Acids

Bile acids are steroids produced by vertebrate biliary system which function as detergents to solubilize ingested fats which are then either resorbed or released to the environment. Bile acids, released by young stream resident salmonids, have been hypothesized to function as migratory attractants for returning adults.

Gonadal Steroids

Mature fish release gonadal steroids to function as pheromones. The gold fish has been an important model in these studies, and it is now clear that the gold fish olfactory system is acutely and specifically sensitive to three steroidal hormone products.

Prostaglandins

Prostaglandins are a class of fatty acids which are traditionally associated with a variety of autocrine and paracrine functions in the vertebrate body. In many fishes, however, F prostaglandins also function as a hoemone that stimulates female sexual behavior. These hormonal compounds are also functioning as potent olfactory stimulants with pheromonal activity for many species of fish.

Chemosensory Behavior

Summarizing some of the basic components of chemosensory-mediated behavior of fishes and placing them into five categories; (1) feeding; (2) detecting danger; (3) nonsexual social interaction; (4) orientation and (5) reproductive synchrony, olfaction appears to be involved with all the five behaviors

Feeding Behavior

Chemosensory cues play an extremely important role in food finding and recognition in most species of fishes. Chemical cues are particularly important to benthic living and omnivorous fishes. Although most studies have tested fishes of commercial interest for designing feeding attractants/stimulants for aquaculture and/or harvest, there has been little work on the vast majority of other species or in determining the precise sensory and behavioral mechanisms underlying feeding. Without exception, all studies have found normal feeding behavior to be mediated by mixtures of L-amino acids, with nucleotides and betaine also having a role in some species. Traditional models of fish behavior have supposed that olfaction is responsible for arousal (recognition of food odor at a distance) and searching, while taste sense is responsible for food uptake and ingestion. While this is likely the case for many fish (namely, salmonids), it is not true for fishes with well-developed external taste systems, such as catfishes, which are capable of locating food odors at some distance using taste alone. Another complicating issue is the disparity in the gustatory spectra of different fishes. Some fishes taste a wide range of amino acids (namely, catfish), while others detect just a few (salmonids). While the olfactory systems of most fishes are generally believed to be sensitive to amino acids, it is also clear that the olfactory sensitivities of fishes vary. Thus the extent with which taste and olfactory systems overlap varies.

Learning is one aspect of chemosensory mediated feeding behavior which is clearly associated with olfaction. Recent evidence in the rainbow trout, a species which lacks a well-developed external taste system, indicates that the expression of appetitive behavior in this species requires a functioning olfactory system.

Like other vertebrates, many fishes appear to learn particular odor characteristics of their preferred prey, which is dependent upon the olfactory sense.

Stimuli Associated with Danger

When injured, many species of fish release compounds from their skin which serve as an alarm "pheromone" causing conspecifics to either flee or hide. Species specificity is generally observed, suggesting instinctual recognition. Olfactory ablation has established that this response is olfactory. Many groups of fishes including the Percidae use fright substances. Different cell types and compounds are likely involved in different groups. It has been found that responsiveness to the odor of injured fishes can be learned by fish species that are not taxonomically related to the injured species. In a more complex scenario, juvenile brook char, *Salvelinus fontinalis*, come to avoid the odor of Atlantic salmon if the salmon are fed other brook trout. This response is dependent upon a functioning olfactory sense

Detecting Predator and Prey

Fish use olfactory cues to recognize conspecifics potential mates, offspring and individuals. Water soluble sex hormones released by an individual can produce dramatic changes in the behavior and physiology of receivers. It is interesting to note that male half-naked hatchet fish, *Argyropetecus hemigymnus* (Sternoptychtidae) have the biggest nose relative to body size of any vertebrate. This permitted this bathypelagic fish to locate mates despite their extremely low densities.

In general, very little is known about either signaling behavior or signal structure in fish olfactory communication. A notable exception is chemical alarm cues or *Schreckstoff*.

Fish is capable of detecting their predator and prey using at least four different sensory modalities; electrical, sound (including pressure), chemical and visual.

To chemically detect a predator, it is important that there be a chemical reliably associated with a predator both in space and time. In particular, for fish that occupy relatively small bodies of water, the chemical must either rapidly degrade or diffuse to levels which are not detectable. This is critical because otherwise the correlation between the presence of chemical and the likelihood of encountering a predator will diminish. With a low correlation, detection of the chemical will provide little or no useful information regarding the presence of predators and the risk associated with remaining within a particular habitat.

The best known chemical for detecting predators or recent predation is alarm substance (AS) or *Schreckstoff*. This chemical is possessed by fish within the super order Osatariophysi and is generally agreed to be hypoxanthine 3-N oxide. The specific components that make this chemical biologically active have now been determined. AS resides within the club cells of epidermis of these fish and can only be released when the fish suffered physical damage that rupture these cells (that is, when it has been injured by an attacking predator). Fish that are capable of detecting AS often exhibit a fright reaction that can include a range of specific anti-predator behaviors, such as, evasive swimming maneuvers, reduced motion and movement to different habitats. In the field AS will cause individuals to avoid the area for up to 12 hours after release. Other individuals will move into this location after only 3 hours, demonstrating the detection of the chemical cause, fish to remember a location as being dangerous long after the chemical is no longer present in the environment.

Fish need not only rely upon injured conspecifics to release this chemical. Some predators like pike (*Esox lucius*) will release this chemical after they have consumed an Ostariophysial. The value of this chemical not restricted only to species that produce this pheromone. Recent experiments have demonstrated that species that do not produce AS are capable of detecting and responding this chemical, including some salmonids and sticklebacks.

Crucian carp (*Carassius carassius*) are also able to chemically detect their predators, although their response to this information is not restricted only to anti-predator behavior. Patterson *et.al.* (2000) have demonstrated that when carp detect the odor of a predator that has consumed prey containing AS, they respond by modifying their pattern of growth. In the presence of these cues, individuals from populations that contain predators will alter their pattern of growth so that they become deeper bodied.

Of more importance to prey are tactics to allow them to escape predation, and specifically to defeat predators by employing the most effective technique of all- not fitting in their mouth. Relationships between gape limitation (the biggest possible prey that can be consumed) and body size have been characterized for several species. The anti-predator benefits of larger body size are amplified further when recognizing that predators have also predators.

But the deep bodied morph had an impaired ability to compete for food, since that this deep bodied morphology must have a higher drag when they increased their velocity and a significant of energy requirements beyond what was observed for the normal morph, suggesting that they would fare poorly under intense intrasecific competition with the normal morph.

Sticklebacks appear to develop lateral plates and armor in habitats that have greater densities of predators. The combination of erect spines and lateral plates can render these fish immune to smaller predators although this morphology may make them more vulnerable to invertebrate prdators.

AS is not the only chemical known to generate anti-predator behavior. Some salmonids are also known to respond to the presence of the amino acid, L-serine that is commonly found on the skin of mammals. When this chemical is detected, these fish will cease moving and feeding, and it is believed that such a response will then prevent their response to many types of gear associated with sport fishing. Such a response indicates that fishes are capable of detecting and using any information that is reliable in signaling the presence of danger.

Social Interaction

Olfactory cues have important roles mediating non-reproductive social interactions of conspecifics, the nature of which varies with the behavioral repertoire and life histories of the animal in question Some of these roles are certainly pheromonal (associated with instinctual responsiveness to conspecifics), and some are likely learned. Among the non-reproductive roles associated with olfactory function is the recognition of related individuals. Example of individual was described for the bullhead catfish, *Ictalurus nebulosus,* which live communally and appear to recognize dominant individuals on the basis of olfactory cues, of which dietary amino acids appear to be important constituents. Kin recognition has been suggested to be of importance in many species of schooling fish, and examples exist which demonstrate that when closely related salmonids are raised together, they have the ability to recognize kin based on odor alone. On a larger scale, population and/or species-specific odors appear to play a role in migratory orientation of certain anadromous fishes.

Migration and Orientation

Many species of migratory fishes are known for their abilities to find from great distances locations suitable for feeding and/or breeding.These fishes may be divided into two groups; those that "home" or return to specific geographic locations (generally the place they were born), and those that find locations with particular characteristics (namely, good habitat for larval survival). In both instances, it is clear that olfactory cues are largely responsible for their ability to locate particular location. Pacific salmon are classic example of fish that home, and there is strong evidence that they imprint to (instinctually learn over the course of a limited period of time) the odors of their natal streams while living there as juveniles. Salmon without functional olfactory systems have very little success returning to natal streams. Although no progress has been made in identifying the chemical nature of imprinted odors, it has been suggested that a complex mixture of compounds which includes organics from plants as well as conspecific odorants, is involved. The sea lamprey, *Petromyzon marinus,* is a good example of a fish that does not return to a home stream to spawn, but instead selects a stream with a large number of conspecific larvae in which to spawn. It has been hypothesized that conspecific bile acids released by larval migratory fishes may provide a species/population specific pheromone which guides adults back to these streams. The olfactory sense of lamprey has been found to be specifically sensitive to two bile acids, petromyzonol sulfate and alcoholic acid that are unique to larvae, one of which is released in substantial quantities by larvae. Behavioral responses to these compounds have also been noted. The roles of bile acids remains to be demonstrated in teleost fish. The possibility of earthy odors (compounds originating from terrestrial and freshwater microbes) may have a role in the instinctual recognition of fresh water by eels and other catadromous species.

Reproductive Cues

All species of fish when sexually mature have been found to produce cues that signal their condition to conspecifics. Moreover, when examined, sex pheromones have been shown to be detected by the olfactory system. These cues serve a wide variety of function in different species and can be classified as being either "primers", which evoke primarily endocrinological responses, or "releasers", which trigger behavioral changes. Although pheromone production is generally associated with females, examples exist of nest-guarding males which release cues to attract females. Similarly, the importance and use of these cues vary; for instance, in gold fish, sex pheromones are so important that olfactory ablated fishes experience near total reproductive failure. On the other hand, anosmic salmonids merely experience reduced levels of circulating hormones and sperm production.

Extreme olfactory sensitivity of a number of fish species to various gonadal steroids and to their metabolites has now been demonstrated using electroolfactogram recordings (EOG). Although a great variety of steroids have been found to function as olfactory stimulants, the olfactory spectra of individual species is typically restricted to one to five compounds. The sensitivities of different species of fish can be categorized along taxonomic lines with distinct differences in the pattern of compounds detected being consistently evident only at the level of the sub-family or higher. Notable example of other fish species for which both behavioral and olfactory sensitivity to steroids has been shown include the gold fish, African catfish (*Clarius gariepinus*) and Atlantic salmon (*Salmo salar*).

Only for the gold fish has pheromone production, release, olfactory sensitivity and biological responsiveness has been described. Gold fish ovulate in the spring in response to a surge in gonadotropin triggered by rising temperature, aquatic vegetation, and pheromones. Because females spawn in murky waters at daybreak, male-female reproductive physiology and behavior must be tightly synchronized. This synchrony is mediated by at least two hormonally derived cues with distinctly different actions and identities. The first pheromone is released prior to spawning by ovulatory females and functions primarily as steroidal printer. The second pheromone is released by recently ovulated (sexually active) females and stimulates male sexual activity, thus functioning as releaser. It presently appears that both cues are rather common hormonal metabolites, which the male's olfactory system has evolved to detect. The principal function of the preovulatory pheromone appears to be to stimulate increased sperm production and motility in conspecific males and to evoke behavioral competitiveness. The preovulatory pheromone is released by females immediately following ovulation and coincides with increases in circulating PGF which is also responsible for female sexual behavior. It is now clear that the PGF pheromone is comprised of a mixture of 15K-PGF and PGF and dominated by the former. EOG recording has shown that these cues are detected by different olfactory receptor mechanisms and that sensitivity to them is sexually dimorphic.

Chapter 16

Fish Behavior in the Aquarium and in Wild

In the Aquarium

In contrast to the wild environment, the aquarium environment is more stressful to fishes, due to its very limited space, static environment, limitation for growth and threat of un-congenial ecosystem. All the essential components for sustaining a healthy life, growth and reproduction is not only limited in the aquarium, but at times become critical for their lives and they had to struggle hard for their existence Four things are essential for keeping the fishes in good condition in an aquarium.

1. Suitable water
2. Sufficient dissolved oxygen
3. Correct temperature
4. Correct feeding

Water that is supplied to houses in most cities and large communities is harmful to aquarium fishes; goldfish as well as tropical fish. Such water usually contains chlorine to render the water sterile and safe for human consumption. But this chemical is injurious to all fish. Deep well or artesian water is also not suitable since it often contains a large amount of injurious minerals. Highly acid swamp water and fresh rain water is not suitable. Well water that lacks a high mineral content and most clear water from ponds, lakes and streams are satisfactory for aquarium use. As an aid in maintaining clear and healthy water in an aquarium, filters using activated carbon and fiber glass and operated by an electric air pump are often used.

The problem of maintaining sufficient oxygen in the aquarium is closely associated with the possibility of overcrowding the tank with fish, with the temperature of the water, and with the diverse oxygen requirements of the many popular varieties of tropical toy fishes. The water in an aquarium is dependent largely upon the oxygen absorbed from the air at the surface of the tank. It is often recommended that 62.5 square centimeter of surface be allowed for each 2.5 cm fish. Plants in the aquarium are very decorative, and are useful as refuge for fish in the tank but, contrary to general opinion, do not provide fish with the required oxygen. When the fish in an aquarium come to the surface and remain there almost continually, it is usually a sign of insufficient oxygen resulting from overcrowding or overfeeding.

Most aquarium fishes will tolerate a temperature range of about 10 degree F. The usual temperature range for most popular tropical fishes is from 72 to 80 degree F, but goldfish do better at a lower temperature. Sudden changes in temperature are fatal to most fish, regardless of whether the temperature is raised or lowered. A temperature difference of more than 2 degree F should be avoided when transferring fish from one tank to another or during changing aquarium water.

The amount of food eaten by aquarium fishes is largely determined by oxygen content and temperature of the water, as both factors have a direct effect upon fish metabolism or body activity. Usually, the warmer the water (well supplied with oxygen), the greater the movement and greater the hunger of the fish. The fish should not be fed more than they will consume within five minutes. Uneaten food must be removed promptly to prevent fouling of the water and lowering of the oxygen content.

The occupant of the aquarium has to rely on the mercy of the aquarium keeper in providing and changing suitable water, having sufficient dissolved oxygen at an optimum temperature and for correct feeding.

The movement of fish in an aquarium is confined to limitation of space. Unlike in larger and deeper water areas, the fish movements both horizontal and vertical, is confined to the dimension of the aquarium for horizontal swimming and the depth of water maintained in the aquarium for vertical diving. They move round and round horizontally and up and down vertically till they are tired. Since there is no thermal stratification and the movement of water (except during aeration, fishes do not find a temperature difference in water layers and they has to adjust with the temperature of the whole water mass (be it hot or cold).

The aquarium environment do not produce food of its own except some microscopic plants either on the glass walls of the aquarium or on the bottom pebbles and sand, when the unused food and fishes excreta get decomposed in the aquarium water, when water was not replenished for long lime and the aquarium is kept in a well lighted place. But these planktonic plants do not serve either qualitatively or quantitatively the requirement of food for proper growth of the fishes. In an experiment, when 16 new born of black moly of the same stock were reared purely on the food generated in an 18 inches aquarium containing 12 liters of water without any food supply from outside, the young mollies were found to nibble on the algal deposits on the aquarium walls and bottom. But as they grew in size, the food was insufficient and scarce and they were stunted, emaciated and ultimately died one after another from 18th day onwards.

In most of the aquarium, live or dried food are supplied from outside. When the food are put in the aquarium (either in the feeding cup or spread on the water surface, fishes were found to rush towards the food and hurriedly consume and a sort of competition is created among the fishes of the aquarium for grabbing the food as quickly as possible. There remains no scope for the preference of food by different fish species. So the type of food given to the aquarium may be preferred food for some species, but may be emergency food for others for sustaining lives. So the food requirement for non-preferred group is not fulfilled and they suffer from stunted growth due to loss of appetite. Some of them may become sick, emaciated and died. In aquarium, the normal growth of fish is hampered due to stress facors.

Fishes of different species in aquarium learn new behavior by watching or interacting between them quickly due to their close association in a confined space. Compatibility and social harmony have been observed among different fish species (except the predator-prey related fishes) in an aquarium. Except the male Shyamese fighter (*Beta speledens*) other compatible fish species do not fight among each other for food and space in the aquarium. Within the limited space in the aquarium, each species try to occupy a nitche of their own and flocks along with the other members of own species.

A few species, under suitable conditions (suitable water temperature, suitable breeding habitat with aquatic plants and with proper growth) develop gonads and ovulate at optimum conditions, either give birth to young ones (live bearers) or lay eggs, when the proportionate numbers of males of the species are available in the aquarium. In case of live bearers, (guppy, moly, swordtail etc) the males have been found to chase the ovulated females and impregnate them. In case of egg layers (gold fish), females have been found to deposit eggs on water plants and the males to eject milt over the eggs to fertilize them at suitable breeding conditions (suitable water temperature, time of the day, photo-period etc.).

Parental care by the way of nest making with bubbles, leaves and watching the developing embryos have been noticed in some of the fish species (*Trichogaster*, gold fish etc.).

In the Wild

In the wild fishes do live in a variety of habitat from lotic (moving waters, namely, rivers, streams, estuaries etc.) to lentic waters (stagnant waters, like, lake, reservoirs, wet lands etc.) from freshwaters (rivers, lakes streams) to saline water (seas and oceans) and also in between a mixture of saline and freshwater, namely, estuaries, lagoons etc.Besides these broad divisions of the habitat depending on the physical movement of water masses and interaction with the abiotic environment, each of the division has several micro-habitat (depending on the velocity of moving water, their locations at different altitudes and physical underwater structures and vegetation). Each of the habitat has its own speciality and fishes have to adjust and acclimatize to thrive in the environment.

In order to understand the behavior of fish in wild, it is necessary to know about the organisms as well as the environment in which the organisms live. There exists inter-relation between the organisms and the environmental factors.

Inland waters are landlocked bodies of water and are mainly classified on the basis of drainage basins, each basin constituting a discrete and isolated unit of environment. The chemistry of inland waters are dominated by the composition of rocks and soils of the basin and drainage system and therefore, may vary from basin to basin, if small, and from one part of the basin to another part if it is large.

Oceans are rather continuous and interchange of water from one part to other parts takes place. The total concentration of salt present may vary from place to place and time to time, but relative composition remains more or less constant. The total salt content of ocean waters is about 150 to 200 times that of freshwater.

On the other hand, estuarine or brackish waters are the result of mixing between the oceanic salty water and that of inland freshwater, and the chemistry of which would be determined by mixing proportions of these waters. The biology of estuarine waters is also intermediate those of oceanic and inland waters.

Flowing or Running Waters (Lotic Series, Namely, Creeks, Brooks, Streams, Rivers)

In running waters there is a continuous current only in one direction, "new" water continually coming from source region. Current velocity is a prominent feature of this type of inland water wherein the fishes show specific adaptation.

Certain progressive, inevitable changes occur in the units of this series and one unit may give rise to the other unit in a definite order after a lapse of time. A tiny rivulet gradually deepens, widens and cuts bank at its head, thus continuously enlarging in size and form with time and may change into a brook, which may finally result in a stream or river, namely, order of evolution as follows; Rivulet → Brook → Stream → River. Thus it would be observed that in the evolution of this series conditions continuously change, resulting the migration of environment. Head water conditions migrate further and further inland followed by succession of environmental characteristics of brooks, streams, and rivers. Therefore, fishes characteristics of a particular environment must accompany these migrations, or become adapted to the gradually changing condition, or become extinct. However, these changes are very slow and gradual so that necessary migration or adaptation could be easily carried out.

Standing Waters

Lakes, ponds, bogs, swamps for the lentic series. Current is not a prominent feature of this type of water. Instead, vertical temperature, density and chemical stratification is the dominant characteristic.

The progressive, inevitable changes, occurring in the waters of lentic series, is of filling up of the basins, which they occupy. The changes takes place according to the order as, Lake → Pond → Swamp. Thus the ultimate fate of the lentic series is that it becomes finally extinct. The time required for changes in the units of lentic series from one another depends mostly on; the size of the unit, and the agencies responsible for bringing this change among which are listed, wind, incoming streams, wave action and erosion, plants, animals etc. These definite changes occurring in the lentic series in stages have profound influence on the fate and history of the standing water fishes.

Adaptation and Behavior of Fishes

The head waters in the riversystems in the high altitudes are characterized by high transparency, high dissolved oxygen content and sparse biota. Fishes of these reaches are small sized with special organs of attachment to enable them to live in fast turbulent cold streams. They are Mahseer, Snow trouts, Barilius and minor carps.

The warm water zone of plains with deep pools in the silt covered sandy bed is characterized by high turbidity and detritus loading resulting adequate dissolved oxygen in the surface layer, but meager in the bottom. The region provides the richest capture fishery with larger fishes and giant prawns of commercial importance (Major carps, Other carps, Clupeoids, larger catfishes, Featherbacks and freshwater prawns).

The deltaic reaches, the transition zone between freshwater and salt water occupies the marshy deltaic mangrove forests. The fish fauna of this zone consists of oligohaline (residents of river and can not tolerate more than 0.1 per cent salinity), estuarine (resides in estuaries and can tolerate wide range of salinity), eurihaline (inhabits from sea to upper reaches of estuary can tolerate 1.5 per cent salinity) and stenohaline (living in sea and at mouths of estuary up to 0.12 per cent salinity). It is one of the more productive zones of the river basin and contribute to important diadromous fisheries.

Marine Environment and Ecosystems

The ocean comprise the largest of all biospheres or regions of life on our planet, being some 300 times greater by volume than the living space on and over the land. The environment of the sea is divided ecologically, according to depth, distance from the land and degree of light penetration, in general, characteristic type of organisms occur in these various environmental divisions.

The primary division is the overlaying water mass, the pelagic zone and the land mass beneath it, the benthic zone. The pelagic realm is subdivided horizontally into a neretic province and an oceanic province. The neretic province is composed of all the waters above the continental shelf extending offshore to a depth of 200 m. The oceanic province consists of all the waters beyond the continental shelf with depth greater than 200 m.

The neretic province, being contiguous with the continent and land mass, is a well lighted but a turbulent section of the ocean. Seasonal variation coupled with the influence of moon characterizes the area by strong wave action, marked currents and broad changes in temperature, salinity and nutrients. Dissolved oxygen is high. Detritus in suspension plus extensive micro and macro-biological production imparts a characteristic green, grey, brown or red color to these near shore waters. Representative waters include waters of shore line, kelp beds, coral or rock reefs etc. The number of potential habitats is thus large and varied.

The oceanic province extends from pole to pole, from one continent shelf to another and from surface to the greatest depth. Seasonal fluctuations influence only limited portions of this section, mainly the upper layers. The remainder of this province if fairly stable, uniform and with relatively few habitat types. Productivity is markedly less than the neretic province. Vertical divisions of the province are;

The epipelagic zone (0-200 m) of the oceanic province is characteristically blue in color with good light penetration. The cool nutrient rich waters provide the basis for exceptional planktonic growth, which in turn, supports substantial population of larger animals.

The mesopelagic zone (200-1000 m) is characterized by decreasing light intensity, predominantly in the blues, decreasing temperatures and dissolved oxygen and steadily increasing water pressure. In general, conditions are relatively stable.

The bathypelagic zone is characterized by darkness, cold, high pressure and biological activity. Seasonal variations are essentially nil.

The abyssal area (4000 m + m) of the ocean extends over half of the earth's outer perimeter. It is characterized by darkness, high pressure (200-1000 atmos.), cold (less than 4 degree Celsius) and low levels of dissolved oxygen. Physical change or variation appears slight, if any.

Mangrove Ecosystems

Mangroves is a general term used to describe a variety of tropical inshore communities dominated by several species of trees or shrubs, that grow in salt water. Between latitudes 30 degree N and 30 degree S, the shoreline marsh vegetation is replaced by a community of mangroves.

Mangroves are the breeding and nursery grounds for several fish species. About 400 species of fish are reported to depend upon mangrove habitat. Besides many other fish species visit the mangrove environment, some frequently and others occasionally. Some common species are scats, milk fish, mud skippers, mullets, catfishes, perches etc. Mudskippers live on the mud flats associated with mangrove shores. These are fishes well adapted to alternating period of exposure to air and submersion and are seen frequently hopping along the mud at the water's edge. Thus respire under water like other fishes, but when out of water they gulp air. When submerged, they swim like other fishes, but on land they move resorting to a series of skips. When a mudskipper is out of water, it carries in its expanded gill chamber a reserve of air, from which they extract oxygen. After a few minutes, when the reserve is exhausted, it is replenished from pool or from water in the burrows, which they dig.

Coral Reef Ecosystem

Coral reefs form the most dynamic ecosystem providing shelter and nourishment to many thousands marine flora and fauna. They are the protectors of the coast line. This unique ecosystem is most productive because of its ability to retain and recycle nutrient elements within the ecosystem as well as within animal-plant associations.

Over 1000 of fish are found in the coral reef ecosystem. They includes groups like, damsel fishes, butterfly fishes, sweet lips, angel fishes, parrot fishes, snappers, wrasses, groupers and surgeon fishes. Another 20 per cent are cryptic and nocturnal species that are confined primarily to caverns and reef crevices during day light. The assemblage includes families, such as, the cusk eels, some groupers and their relatives, most of the moray eels and some scorpion fishes, wrasses and nocturnal families including the squirrel fishes, cardinal fishes and sweet lips. Another 10 per cent of fishes including snake eels, worm eels, various rays, lizard fishes, grab fishes, flat fishes, some wrasses and gobies dwell primarily on reefs covered with sand and rubble. A relatively small 5 per cent of the fauna is transient mid-water reef species that roam over large areas. This group includes most sharks, jacks, fusiliers, barracudas and scattering representatives of other famiies

Reef Fishes

The species richness of coral reef fishes is similar to that in corals. Coral reefs harbor more species and diverse fish communities than any environment on earth. In the Great Barrier Reef system, a single reef contain 500 species of fish.

The high diversity of species on reefs is due to the great variety of habitat that exists on reefs. Reefs, areas of sand, various caves and crevices, areas of algae, shallow and deep water and different zones progressing across the reef offers habitat diversity.

Four opposing theories of reef fish diversity and community structure have arisen to explain high diversity of coral reef fishes, particularly in local areas. According to the competition model, the high diversity is the result of strong competitive interactions following recruitment that lead to a high degree of specialization. Each species has a specific set of adaptations that give it the competitive edge in at least one situation on a reef. That is to say, these fishes have narrower ecological niches, so more species can be accommodated in a given area. It is partially supported by correlations between fish species diversity and habitat complexity. The second model rests on fact that most coral reef fishes produce large numbers of larvae that are dispersed into the plankton. Fishes are not specialized (many similar species have the same requirements) and there is active competition among species. Local success and persistence result from chance as to which species of the planktonic larval pool occupies the vacant space. Here competition is unimportant and recruitment dominates. The predation-disturbance model promulgate that the fish populations do not reach equilibrium. Predation, catastrophe, and unpredictable recruitment ensure that populations never become large enough to undergo competitive exclusion, as they are kept below the numbers at which resources or food becoming limiting. According to the most recent recruitment limitation model the larval supply is never sufficient for the adult population size to reach the carrying capacity. The adult population reflects variation in the larval recruitment and not post recruitment events and the fish recruitment is variable. Most reef fishes, despite their obvious mobility are restricted to certain areas of the reef and are very localized, essentially sedentary.

Most fishes do not migrate, and many of the smaller species, such as, gobies, blennies and damselfishes defend territories. But certain larger fish do migrate for feeding areas to a distance of few meters to several kilometers and are often predictable.

Coral reef fishes differ between day and night. The majority of fish species are visible in day. At night, these diurnal fishes seek shelter in the reef and are replaced by a smaller number of nocturnal species not seen during the day. Since some of the nocturnal species are ecologically similar to certain diurnal species (Apogonidae replace Pomacentridae), this is another way of permitting a greater number of species to exist on the reef without competing directly. All nocturnal species of reef fishes are oredaceous, but the diurnal fishes span nearly all trophic categories (carnivores, planktivores, omnivores etc.).

Carnivores constitute 50 to 70 per cent of coral reef fishes. They are opportunistic taking what ever is available to them. They also feed on different prey at different stages in their life cycles. Some specialized carnivores exist. The number of true scavenger fishes are few as the carnivores also act as scavengers and pick up any recently dead organisms.

Herbivores and coral grazers make up the next largest group of fishes (about 15 per cent of the species) and the most important of these are the families Scaridae and Acanthuridae. The remaining fishes are omnivorous and include representatives from virtually all families of fishes on the reef (Pomacentridae, Chaetodontidae, Pomacanthidae, Monocanthidae, Ostraciontidae, Tetraodontidae). Only a few fishes are zooplankton feeders and they are mainly small schooling fishes of the families Pomacentridae, Clupeidae and Atherinidae. Feeding of some fishes such as grunts (haemulids) has side effect of enhancing the nutrition and growth of corals. Schools of certain fishes, which rest by day in coral heads, feed at night on seagrass beds and then return to defecate material rich in nitrogen and phosphorus into the coral heads. The result is faster coral growth.

Many of the fishes produce toxic substances. This may be venom associated with spines or poisonous material extruded onto the body surface (crinotoxin) or the flesh or internal organ may be toxic. Truly venomous fishes are rare on reefs, but a large number of fishes have toxic secretions on their body surfaces (parrot fish, wrasses and surgeon fish to deter predation by abundant carnivores. Toxic flesh or internal organ of fish has caused serious disease (ciguatera) among humans by eating tropical fishes with toxins in the flesh or organs.

Color is an outstanding characteristics of coral reef fishes. The bright warning coloration seems to advertise that the species is toxic or otherwise distasteful and predators will avoid them. In addition to color role in various predator-prey interactions, the color and pattern may serve in species and gender recognition and be used in courtship and mating behaviors.

Cleaning behaviors of some reef fishes is a specialized form of predation in which certain small fishes (Labroides species) or shrimps remove various ectoparasites from larger fishes. This cleaning behavior is widespread and occurs in all reefs. The cleaner fishes or shrimps often set up "cleaning station", where they advertise their presence through bright contrasting colors. The fish to be cleaned comes to the cleaning station area (often a prominent coral head or boulder) and remain motionless as the cleaner moves over its body removing the parasites. The cleaners may even enter the mouth and gill chambers of the fish. Fishes even "line up" at these stations awaiting their turn for cleaning. It is believed that if cleaners are removed from a reef, the fish fauna dectreases, moving away or otherwise showing signs of distress.

The reef fish, *Myripristis murdjan*, feeds primarily on megalopa, other decapods and stomatopods. It is a nocturnal benthic feeder, which preys on or just above the substrate, rising into the water column 35 to 50 minutes after sunset. At 50 to 30 minutes before sunrise it seeks shelter again. Differences in feeding behavior can be observed between the places. The reefs with poor coral development have different feeding pattern than with places with a greater coral variety. In places, with the occurrence of plankton patches, mysids dominate in the stomachs of *M. murdjan*.

Red mullets (Mullidae), feeding on sand and mud flats was confirmed for *Mulloidichthys auriflamma* by examining stomach contents and underwater observations. Food preferences differ between locations. In one location *M. auriflamma* prey more on Brachyura and some stomatopoda whereas in other place Tanaidacea dominate the stomach contents, among a great number of animal groups. Two other species of this group, namely, *Mulloides vanicolensis* and *Mulloides flavolineatus* obtained from other locations feeds xanthid crabs and Polychaeta respectively.

The genus *Caesio,* which includes seven species is divided into two groups. The yellow tail group, of which *C. cuning* is the most common species, consists of three species. *C. cuning* preys mostly on zooplankton. The second group of which *C. chrysozonus, C. lunaris* and *C. coerulaureus* are common in Indo Pacific, prey mostly on larger copepod species such as, *Undinula, Euchaeta, Eucalanus, Candacia, Labidocera*n and on Chaetognatha, Polychaeta, crustacean larvae and fish larvae. A great diversity in the feeding pattern of *C. coerulaureus* was observed.

Mangroves and Coral Reef Fish Communities

Mangrove forests are one of the world's most threatened tropical ecosystems with global loss exceeding 35 per cent. Juvenile coral reef fish often inhabit mangroves. The mangroves are important, serving as an intermediate nursery habitat that may increase the survivorship of young fish. Mangroves in the Caribbean strongly influence the community structure of fish on neighbouring coral reefs. In addition, the biomass of several commercially important species is more than doubled when adult habitat is connected to mangroves. The largest herbivorous fish in the Atlantic, *Scarus guacamaia* has a functional dependency on mangroves and has suffered local extinction after mangrove removal. It was surprising to find that mangrove extent was a dominant factor in structuring reef fish communities. Mangrove extent not only explained a significant component of community structure, but it usually exceeded the influence of reef systems. The interactions within the fish community are so strong enough that mangrove deforestation will also affect populations of obligate reef species.

Mangroves may enhance adult fish biomass in two ways. First efflux of detritus and nutrients may enrich primary production in the neighbouring ecosystems. Second, mangrove nurseries may provide a refuge from predators and plentiful food that increases the survivorship of juveniles. The size frequency of *Haemulon sciurus* suggests an ontogenetic shift in habitat use from sea grass to mangroves, to patch reefs and finally to forereefs, their main adult habitat. Data shows that juvenile grunts migrate from seagrass beds when they reach a length of 4-6 cm.

Migration occurs from seagrass to mangroves, but if mangroves are absent, the grunts move to reefs. Because mangroves offer refuge and the biomass of haemulid predators is greater on reefs than in mangroves, the chances of grunt survival may be lower if grunts migrate directly to reefs. In short, some fish species move to their adult habitat in stages. As the biomass of predators increases at each stage, it is desirable to grow as large as possible before taking the next step towards adult habitat. Mangroves provide an intermediate nursery stage between seagrass beds and patch reefs, and they therefore alleviate a predatory bottleneck in early demersal ontogeny.

Visual Communication in Tropical Coral Reef Fish

The reef habitat is one of the most optically complex environments. Coral reef fish are renowned for their distinctive color patterns, especially those belonging to the families Acanthuridae, Chaetodontidae, Labridae and Scaridae. Functional significance of reef fish color patterns was thought of for antipredation disruptive coloration, territory advertisement, species recognition and intraspecific coordination of schooling and feeding. But the basic assumption is that the color patterns of fish are conspicuous in their natural habitat.

Conspicuousness can be evaluated by measuring the amount of contrast provided by a color pattern. Contrast is broadly defined as the proportional difference between target and background radiance and two types are considered visually relevant. Luminance contrast refers to differences in brightness resulting from amplitude differences between the spectra. Spectral contrast refers to color differences due to differences in spectral composition, that is, the shape of the spectral distribution.

Detection of color pattern depends upon three processes, (i) light reflectance and transmission to the eye, (ii) light transmission, refraction and photoreception within the eye and (iii) neural mechanisms in the retina and brain.

Light reaching the eye from a color pattern is a function of the ambient light striking the color patch, the patch's reflectance spectrum and the transmission properties of the medium.

Inshore and offshore seas have wide range of photic conditions. The optical characteristics of inshore seas have dominant wavelength correspond to the green region of the spectrum and attenuation is high. The photic conditions are related to high amounts of suspended particles and organic matter due to the close proximity to land.

Photic conditions at the offshore seas are consistent with ocean water type. The spectral distribution of light is shifted towards the blue part of the spectrum and the attenuation is relatively low.

Underwater light conditions varied as a function of direction. In the downwelling direction irradiance spectra are broad, while upwelling and horizontal light are shifted to shorter wavelength, which is typical of clear ocean water.

Inherent luminance contrast was highest for *Labroides phthirophagus* while *Chaetodon auriga* has the highest inherent spectral contrast. High inherent luminance contrast of *L. phthirophagus* may have resulted from the yellow head patch next to dark black stripe. When a color pattern has light patch next to the dark patches, it is conspicuous in almost all light conditions, even to the animals without color vision, because it is based only on brightness differences.

Against water backgrounds, the luminance contrast from fish is influenced more byphotic differences between locations and between depths.

Most theories for the adaptive significance of reef fish colors state that colour patterns are conspicuous and they are important in visual communication. The effectiveness of a color pattern as a visual signal differs according to ambient photic conditions which are in turn, influenced by a series of factors, such as, time of day, season, weather and location etc. Therefore, to be effective optical signals, color patterns should be optimized for a particular function and to communicate under a wide variety of photic conditions.Conspicuousness of reef fish depends not only on the color pattern itself, but also on the optical properties of impinging light and background conditions. The physical mechanisms involved in signal generation and visual communication in tropical coral reef fish is understood to some extent from the study of four species of Hawaiian coral fish, namely, *Chaetodon auriga, Labroids phthirophagus, Thalassoma duperrey* and *Zebrasoma flavescens.*

Chapter 17

Diversity and Adaptation in Fish Behavior

A fish according to one dictionary is "an aquatic cold-blooded water breathing, gilled vertebrate, with limbs represented by fins.

It is difficult to describe a typical fish due to their numerous diversity and adaptability. In fact, more than half of all species of vertebrates (back boned animals) are fishes. They have adapted themselves to a wide range of environments. They are found in the icy waters of the polar regions on the one hand; while they exist miraculously uncooked in the hot desert pools up to a temperature well above 100 degree F, on the other hand. They may roam widely over the vast expanses of open sea or spend their entire life in the cramped underground quarters of an artesian well. They thrive in high mountain lakes and also in the abyssal depths of the ocean. They may even leave temporarily the aquatic habitat to scamper over mud flats or climb small trees in search of food. If the pools dry up, they may bury themselves in the mud and spend the dry season, breathing air.

On the whole, it can be said that where there is water, there are fishes; and three fourth of the earth's surface is covered with water.

The size and structures of fishes also follow an enormous diversity as they have adapted to live under wide range of conditions.

A full grown Philippine gobies is only a half inch or less, while a full grown shark attains a length of 40 to 50 feet and some time may reach 70 feet.

Shapes are equally variable; the elongated eel with snake like body; the skates and rays with flattened body; the ocean sunfish with a body as deep as it is long; the globular puffer; the flounders and soles with both eyes on the same side of the head;

and even the popular sea horse, which at first glance would hardly be detected as a fish.

Diversity of Fishes in Wetlands

The Indian fish fauna is divided into two classes, namely, Chondrichthyes and Osteichthyes. The Chondrichthyes are represented by 131 species under 67 genera, 28 families and 10 orders in the Indian region. The Indian Osteichthyes are represented by 2415 species belonging to 902 genera, 226 families and 30 orders, of which five families, notably the family Parapsilorhynchidae is endemic to India. These small hill stream fishes include a single genus, namely, *Parapsilorhynchus*, which contain three species. They occur in the Western Ghats, Satpura Mountains and the Bailadila range of Madhya Predesh only. The fishes of the family Psilorhynchidae with the only genus *Psilorhynchus* are also endemic to the Indian region. Other fishes endemic to India include the genus *Olytra* and the species *Horaichthys setnai* belonging to the families Olyridae and Horaichthyidae respectively. The latter occur from the Gulf of Kutch to the Trivandrum coast. The endemic families form 2.21 per cent of the total bony fish families of the Indian region. 223 endemic fish species are found in India, representing 8.75 per cent of the total fish species known from the Indian region and 128 monotypic genera of fishes found in India, representing 13.20 per cent of the genera of fishes known from the Indian region.

Fish Diversity of Ganga River

Among the vertebrate fauna, though the number of fish species is high (378 species), the population of most of the economically important fishes, including the Indian Major Carps (*Labeo rohita, Catla catla* and *Cirrhinus mrigala*), has dwindled in recent years. But the biomass of the catfishes has been found to increasing. Altogether 75 species of fish were identified to have high commercial value. During a fish stock assessment exercise in the Ganga in a stretch of about 30 km in and around Patna between July 1994 and June 1995, one hundred six species of fish fauna were collected and identified from the river. *Bagarius yarrellii, Silonia silondia* and *Anguila* sp. were found to be the most vulnerable and rare. The degradation and lossof habitat in the river ganga have led to decline of the major carp population while less economic fishes (minor carps and small catfishes) are increasing in relative abundance. The migratory fishes like *Hilsa* are also mainly confined down stream of the Farakka Barrage. After construction of the Barrage, the fishery of *Hilsa* collapsed; however, in recent years their catch was recorded up to Allahabad. Boulder mining and, dams/barrages resulting in water flow regulation in upper reaches of Ganga have also adversely affected the Mahasheer population. Still the fishes from the largest group of living natural resources in the river Ganga and serve as the largest source of fish spawn in India.

Marine Fish Diversity

The marine ecosystem has a varying profile. The coastline encompasses almost all types of intertidal habitat, from hyper saline and brackish water lagoons, estuaries, and coastal marsh and mudflats, to sandy and rocky shores. The subtidal habitats are equally diverse. Each local habitat reflects prevailing environmental factors and is further characterized by its biota. Thus, marine fauna itself demonstrates gradients of change throughout the Indian coasts.

Free swimmers or nekton are important components of marine biodiversity and constitute important fisheries of the world. The dominant taxa in the nekton are fish, others being crustaceans, mollusks, reptiles and mammals. Out of total 22000 finfish species, about 4000 species occur in the Indian Ocean of which 1800 species are reported in the Indian seas. A majority of the fish species is found in the coastal waters. It is estimated that 40 species of sharks and 250 species of bony fishes represented the oceanic species.

Coral Reef Associated Fishes

Fishes comprise about half the total number of vertebrates. The number of estimated living fish species might be close to 28000 in the world. Day (1889) has described 1418 species of fish under 342 genera from British India. Talwar (1991) has described 2546 species of fish belonging to 969 genera, 254 families and 40 orders. The distribution of marine fishes is rather wide and some genera are common to the Indo-Pacific and the Atlantic regions. 57 per cent of the Indian marine fish genera are common to the Indian Seas and to the Atlantic and Mediterranean. The exact number of species associated with coral reefs of India is still to be found, however the number of fishes in the Indian Ocean is 1367. The Lakshadweep Islands have a total of 603 species of fish, about 750 species are found in the Andaman and Nicobar Islands and in Gulf of Mannar Biosphere Reserve is 538. the category of fishes occurring in coral reef ecosystem includes groups, such as, the damselfishes, butterfly fishes, triggerfishes, filefishes, puffers, snappers, hawk fishes, triple fins and most of the wrasses, groupers and gobies. Another 20 per cent are composed of cryptic and nocturnal species that are confined primarily to caverns and reef crevices during day light periods. This assemblage includes such families as the cusk eels, some groupers and their relatives, most of the moray eels and some scorpion fishes, wrasses and nocturnal families including squirrelfishes, cardinal fishes and sweet lips. Another 10 per cent of fishes dwell primarily on reefs covered with sand and rubble including snake eels, worm eels, various rays, lizard fishes, grab fishes, flat fishes and some wrasses and gobies. A relatively small percentage (about 5 per cent) of the fauna is composed of transient and midwater reef species that roam over large areas. This group includes most sharks, jacks, fusiliers, barracudas and a scattering representatives of other families.

Diversity in Freshwater Ecosystem

The freshwater ecosystems encompass a wide spectrum of habitats covering both lentic and lotic water bodies. The former include either temporary or permanent ponds, lakes, floodplain marshes and swamps, while the latter relate to river and stream. The third kind of freshwater ecosystem is now known as wetlands, which are transitional habitats spatially, sandwiched between the land and lentic or lotic ecosystems. Under the freshwater systems, wetlands include a wide variety ranging from temporary ponds to extensive floodplains of large rivers as well as rice fields.

In freshwater ecosystems there are some distinctly aquatic animals like fish; others live atleast for a part of their life in water and still some others that live on land or trees or both (Kingfishers) and depend on aquatic ecosystems or wetlands for fish and other aquatic organisms as their food.

223 fishes mostly belonging to Cypriniformes, Silluriformes and Cyprinodontiformes are endemic to India.

The riverine and freshwater floodplains ecosystems are highly sensitive to any alteration or degradation of water quality. The Ganga river system, which is the original habitat of Indian major carps and spawn, shows loss of diversity, especially at the location of industrial and sewage outfalls.Today, indigenous fish population is threatened in Indian rivers as well as freshwater wetlands. Dams and barrages on river routes obstruct the migration of fishes. Freshwater eels and other catadromous fishes are known to undertake long migration into the deep sea for breeding. The offspring again ascend the river to return home. But barrages prevent adult eels from migrating to the sea for breeding, while the offspring fail to find the homeward route. The Farakka Barrage is one such example of a hydraulic structure erected by man on the river Hoogly. It is possible that due to such structures migration of *Macrobrachium rosenbergii*, which lives in the river and migrates to the estuary for breeding, is hampered/obstructed. As a result, recruitment failure in the prawn population is obvious. Besides the pollution and other effects, over-exploitation of fish and juveniles from the river systems is also recognized as one of the root causes of loss of aquatic biota from the river and freshwater wetlands in India.

Chapter 18

Cooperative Behavior in Fishes: Group Living and Social Network

Learning in Fishes

Like other animals that live in the complex and variable natural world, fish rely on their ability to change and adapt their behavior through learning and memory. It is now recognized that fish can learn and remember different types of information. For instance, they can reliably locate places to forage, react quickly and appropriately when threatened by a predator and identify potential partners for reproduction. Similarly they can assess competitors and can communicate and recognize dominant and subordinate status, which helps prevent conflicts escalating into potentially harmful fights.

Fishes can learn about their environment through a series of trial and error processes; alternatively they can learn to adapt their behavior by observing the behavioral responses of others within that environment (social learning). To survive from one day to the next, the fish rely on their ability to generate decisions that produce appropriate behavioral responses. How finely tuned these decisions are will, in effect, determine the cognitive capacity of that individual.

Learning and Memory

Fish have been shown to discover foraging patches by individual exploration or by observing the behavior of other foragers. A key concept associated with foraging theory hinges on the ability of the animal to assess food patch profitability and to adjust their behavior accordingly. Foraging theory has used Charnov's (1976) Marginal Value Theorem to make predictions about how long an animal will stay at one food patch before moving to another. This predicts that food patches, regardless of their original profitability, should be exploited to a certain critical level, at which the rate of energy gain is equal to the average rate of gain across all feeding patches in the environment. Thus to make a correct decision to remain or leave for an alternative patch, the animal needs to retain a memory of the profitability of previously encountered patches. Such a behavioral strategy allows the animal to forage at optimal rate within its current environment. There is some evidence that fish do learn and remember this type of information; for example, bluegill sunfish (*Lepomis macrochirus*) use prior experience of patch profitability to influence how long they spend in a particular food patch.

An ability to learn also plays a role in how quickly individuals done their handling skills for different types of prey. Where a fish's ability at attacking, manipulating and ingesting prey is quantified over a series of repeated trials, the fish's performance generate a typical learning curve. The fish are initially slow to consume novel prey, but over a few successive trials, their ability to locate and ingest the prey improves until it reaches a point where it can not consume the prey any faster. It has been demonstrated that 15 spined sticklebacks (*Spinachia spinachia*) needed between 5 and 8 trials to learn how to handle a new type of prey. Similar number of trials were also required by bluegill sunfish as they learned how to consume *Daphnia*. This type of learning occurs over fairly short periods of time.

Fish can track changes in food availability in different ways and may rely on a range of learning and memory systems to help them achieve this.They can learn and remember cues or land marks associated with particular food patches, or they can learn to use the foraging behavior of other fishes around them. The duration of the memories associated with these different variables should be tuned to the environment in which a fish find itself. In a frequently changing environment, the value of a memory may be relatively low, because, within a relatively short space of time, the memorized information may no longer be relevant. Thus in a hghly heterogeneous environment, memory duration would be expected to be short and a fish would be expected to continually up date its memory through learning. In contrast, in a stable undisturbed environment memories would be predicted to be longer term. An example illustrating that these predictions are met can be seen in the way that sticklebacks forget the handling skills needed to capture and consume different types of prey. In a comparison of the memory characteristics of three different populations of sticklebacks, Mackney and Hughes (1995) showed that that three spined sticklebacks (*Gasterosteus aculeatus*) sampled from a pond remembered specific prey handling skills for 25 days; however the same species, but an anadromous population from an estuary was found to have a decreased memory duration of only 10 days. Similarly, a short 8-day memory was observed in a population of marine 15 spined sticklebacks (*Spinachia spinachia*). The difference in memory duration appear to reflect the changeable nature of the environment that these different populations experience, from the most stable (the pond population) to very variable (the anadromous and marine populations). Interestingly all the three groups learned how to handle the prey at similar rates, it was only the duration of memory for these skills that differed.

The environmental enrichments promotes cognitive capacity and behavioral flexibility. Evidences suggest that environmental enrichment may have similar effects in hatchery/aquarium reared fish.

A juvenile cod (*Gadus morhua*) reared with experience of variable food and spatial cues are significantly faster at investigating and consuming live prey than cod from same brood stock reared in standard, constant, unchanging aquarium or hatchery conditions. These differences arise even though both groups have been fed on a standard diet of food pellets. Thus introducing some variation into the standard rearing environment generates fish that are more efficient at the pellet-to-live food transition.

Recognize Conspecifics

Being able to recognize other individuals, and to associate certain behaviors with those individuals could be beneficial for a number of reasons. There are a variety of ways in which conspecifics may be recognized. There is good evidences that some species of fish can recognize and preferentially associate with specific individuals that are familiar. It is also possible that fish learn to discriminate between fishusing factors other than individual recognition. Fish can discriminate between general competitive skills. European minnows (*Phoxinus phoxinus*) have been found to prefer to associate with individuals that are poor competitors. To do this, the minnows must be able to discriminate between good and poor competitors. It is not necessarily individual competitive abilities that are recognized and remembered, but rather the overall competitive ability associated with that group or school. The advantage for the minnows to discriminate between good and poor competitors is that a fish feeding on a small limited food patch has a better chance of securing more food if it is surrounded by poor competitors.

Recognition of particular individuals can be important when repeated interactions occur. Sticklebacks prefer to inspect a predator with a partner that they are familiar with and perceive to be cooperative. It has been found that the levels of aggression between sticklebacks that jointly forage on the same resource varies depending on how much the sticklebacks have interacted. As the fish become more familiar with each other, they are less aggressive and more cooperative in foraging task. The effect of familiarity increased over a four week period; after two weeks the fish showed intermediate levels of aggression, but after four weeks the fish showed almost no aggression. Similarly the breakdown of the individual recognition also showed a time-lag response with the sticklebacks showing some recognition after two weeks, but apparently no recognition after four weeks.

Social Learning and Behavior

Social learning is a specialized form of learning where an animal acquires a new behavior by watching or interacting with other animals. At one time, social learning was believed to be so cognitively demanding that it was only to be found in intelligent species. As such fish was largely overlooked, however, there are now a number of empirical field and laboratory studies demonstrating that fish can exhibit competent social learning.

Social learning in fish has been investigated for a range of behaviors from mate choice to foraging. Antipredator behavior clearly has a learned component to it. Learning how best to react in the presence of predator is an excellent example where trial and error learning is extremely expensive; make a wrong decision and it may be last.

The schooling tendency of various fish species may help mediate rapid antipredator responses. Here the three sensory systems that are likely to play key roles, in rapid responses are vision, the lateral line system, and olfaction. Observing that a close neighbor is responding to seeing a predator, detecting recently released alarm substances, or picking up information on a sudden change in the movement of nearby fish could quickly allow individual fish to adjust their own behavior. As a school of fish reacts to the presence of a predator on repeated occasions, the individual within the school can learn from those around them how to respond in a coordinated fashion. Ultimately this can lead to extraordinary levels of precision swimming that are observed when a school of fish suddenly forms a tight, rapidly moving ball in the presence of a predator. Such antipredator responses appear to confuse the predator and make it difficult for it to track an individual fish. Schooling information could in some situations effectively spread without explicit signaling or complex forms of communication.

Social learning is important for several species of fish that are known to possess chemical alarm cues. When these alarm cues are detected they can promote over antipredator responses in conspecifics. These cues provide reliable information regarding local predation risk. Many of the fish capable of producing alarm substances do not have an innate recognition of predators, but instead they need to learn this information. To do this, they need to pair alarm cues with the visual and/ or chemical cues associated with the predator. Once learned (and they may learn this association after one exposure only) the fish are then primed to react to alarm substances released into the environment by conspecifics.

The way in which some fish learn to move efficiently around an environment to find specific resources is now known to be facilitated by social learning. The mating ground of blue head wrasse (*Thalassoma bifasciatum*) migrate to remain constant across generations. To investigate whether each new generation socially learned the migration and the location of the mating sites, Warner (1988, 1990) removed a complete population and replaced there with another group of fish. These transplanted individuals established new mating grounds, and it was these newly established grounds that were subsequently used by the following generations. Such site fidelity for mating grounds across generations can only be reinforced through social learning.

In another species of reef fish, the French grunt (*Haenlulon flavolineatum*) makes daily migration to feeding areas, and after an initial observation that small groups of grunts appear to be joined on these daily migrations by juveniles. It was investigated whether the juveniles socially learn routes to the feeding areas. By moving juveniles between groups and observing their subsequent swim paths when older group members were removed. Heltman and Schultz (1984) demonstrated that these young fish learn their migration paths by following older group members. Both the French grunt and blue head wrasse examples highlight how the day to day spatial movements of some reef fishes lend themselves well to field studies of social learning.

Laland and Williams (1977, 1998) have brought studies of spatial social learning into the laboratory. They found that "naïve" guppies (*Poecilia reticulate*) were able to find their way to a foraging area by following "demonstrators" that had previously been trained to swim a particular route. After a period of overlap between the knowledgable demonstrators and the naïve observers were able to swim the correct route themselves. This illustrates that the observer guppies had socially learned the route to the food area. Being able to assay this type of social learning within the controlled confines of a laboratory tank has provided an excellent method for addressing a variety of related questions.

Day and colleagues (2001) were able to show that when a food patch is visually isolated from the shoal, smarter groups of fish discover it more quickly than larger groups. However, when the food patch is visible through a transparent barrier, larger groups of fish locate the food faster than smaller groups. This curious result is proposed to reflect the motivation to shoal. Fish that join large shoals are reluctant to leave them as larger shoals provide more protection. However, fish that are members of small shoals are less compelled to stay together, which provide opportunities for individual to move away from the shoal and to find food over wider areas.

There are examples that indicate that some species of fish have the capacity to gauge the reliability of information. Perceiving differences in information is fairly straight forward, but understanding that one piece of information is more reliable than the other requires a higher level of information processing.

Coolen and colleagues (2003) demonstrated sympatrically occurring species of stickleback learn about the relative profitability of feeding patches in different ways. These differences are attributed to variation in responses to predators in their micro-habitat. Nine spined sticklebacks (*Pungitius pungitius*) which have very little body plating to protect them from predators, use information provided by others to assess food patch quality. The nine-spined sticklebacks can learn about patch profitability by watching the behavior of conspecifics and, interestingly also by observing hetero-specific three-spined sticklebacks. Three spined sticklebacks on the other hand, were able to locate food patches by observing other fish (both conspecific and hetero-specific) feeding, but they were unable to detect information on the relative profitability of a patch. The well-developed armor plating of the three-spined sticklebacks presumably places these fish at a lower risk of predation compared to the less well-defended nine-spined sticklebacks. These differences in predator defence may force the nine-spined sticklebacks to be more cautious and individuals that can detect not just the location of a food patch but also its relative profitability before leaving shelter to forage will be at an advantage.

In extension to these initial observations, when compared both species of sticklebacks it has been found that nine-spined sticklebacks adjust their foraging decision making to preferentially use reliable information. In particular, it was discovered that fish were able to discriminate between reliable and unreliable information. The sticklebacks were shown to prefer more recent, more reliable observations of conspecifics foraging at a particular location, even when this meant ignoring information that they had previously obtained when they had sampled the patch for themselves.

Spatial Learning

Many aspects of day to day existence rely on an animal's ability to move between different biologically important areas.Fish are no exception to this and many have the ability to move efficiently through an environment using reference points to guide their movements. This can be achieved in number of ways.

Resources are frequently distributed in a nonrandom way within an environment. An ability to move from one point to another taking the shortest possible route or alternatively a route that will least likely expose a fish to a predator will be favored. In some instances, this is achieved by the animal making a series of body-centered, or egocentric movements or turns. In other cases, the animal will orient itself by keeping a track of its own movement relative to one or more external reference points, and sometimes these may be used to form a map. When moving over much longer distances for example, during seasonal migrations fish may use a compass rather than a map to allow them to maintain a particular course or bearing.

Chapter 19

Applied Fish Behavior

Many of the fish behaviors have applied values, not only for catching fishes from different environments, but also is very much useful for domestication of some species for culture, breeding and seed production in hatcheries and also for management strategy and conservation and rational exploitation.

Feeding Mechanism

Mouth expansion (suction), forward swimming (ram) and manipulation are three mechanisms of prey capture in fishes. Inspite, considerable diversity among the approximately twenty thousand species of fishes, the most common mode of prey capture involves suction feeding. Although the general principles underlying suction feeding are relatively straight forward, several aspects of the feeding system can be modulated in order to alter feeding performance. For example, the rate and magnitude of mouth expansion, the distance between the fish's mouth and prey item, and body movements will influence suction-feeding performance. Jaw protrusion is an additional feature that can enhance suction-feeding performance. Based on recent evidence, intraoral processing (chewing) also appears to be a relatively common feature among fishes. For example, bowfin (*Amia calva*), salmonids (namely *Salmo salar*) and pike (namely, *Esox niger*) all chew prey with their oral jaws. Modern experimental techniques, including sonomicrometry, digital particle image velocimetry (DPIV- a modern computational technique used to visualize the movement of fluids. Neutrally buoyant reflective particles are placed in the water, illuminated using a laser light sheet and are used to measure water movement. The movements of the particles during feeding are recorded using high – speed video), *in vivo* pressure recordings, and computational fluid dynamics, all have propelled our understanding of the biomechanics of feeding in fishes.

The mechanisms underlying suction feeding in fishes are fairly complex. The initial expansive phase is a result of hyoid depression, cranial rotation, jaw protrusion, and jaw opening.

Suction feeding is the most common mode of prey capture. It involves the rapid expansion of oral cavity, which ultimately draws water into the mouth due to pressure gradient formed. The majority of bony fishes employs suction feeding in order to capture prey.

Evolution of Suction Feeding

The invasion of new habitats by early fishes was likely facilitated by the evolution of jaws. The diversity of gnathostomes (jawed fishes) exploded in the Silurian and gnathostomes is currently represented by the condricthyans and bony fishes. Later advances, such as jaw protrusion and pharyngeal jaws, then enhanced the basic plan.

Jaws evolved in fishes over 460 million years ago. Two hypothesis have been put forth explaining the origin of jaws. Both the hypothesis agree that jaws represent the first branchial arch, also termed the mandibular arch. The evidence supporting this includes the fact that most of the elements of the jaws (skeletal elements, muscles etc.) resemble those of other branchial arches. Also both agree that the lower jaw resides in the ancestral position of the mandibular arch, whereas the upper jaw stems from the forward extension (in front of the mouth opening) of the maxillary part of the mandibular arch.

The first hypothesis for the origins of jaws is the neoclassical hypothesis, which asserts that the jaw first evolved for stronger ventilation and then to grasp prey. As ventilation increases, prey would be sucked in and potentially consumed. Thus what originally began with a ventilatory function eventually became utilized for suction feeding, indicating that the current use in feeding is an exceptation. Eventually, the ability to grasp prey was improved when dentition became associated with the mandibular arches. In the neoclassical hypothesis, few or no oral structures were lost during the jawless-to-jawed transition.

The second hypothesis for the evolution of the jaws is the heterotopic hypothesis, which is based mainly on modern developmental studies. The hypothesis asserts that the upper jaw did not evolve through a series of intermediate stages. Instead, certain genes are expressed farther back in the head of gnathostomes compared with lamprey embryos, which leads to different skeletal structures being generated by different parts of the neural crest between the two clades.

Performance of Suction Feeding

In general, generating high fluid speeds (due to negative pressure in the mouth cavity) during suction feeding is a good indication of suction-feeding performance. However, several other factors have been implicated in suction-feeding performance, including the ingested volume of water (IVW), the volumetric flow rate, strike accuracy, and fluid acceleration. The importance of these metrics depends on the ecology of the fish. Large mouth bass (*Micropterus salmoides*) depend strongly on a large ingested volume of water and high forward movement speeds in order to capture prey. For this species, maximum fluid speed is likely not as important for capturing prey. Thus species-dependent metrics of suction feeding performance are necessary to understand the feeding diversity of fishes.

Hydrodynamics of Suction Feeding

Modern techniques have greatly enhanced the understanding of feeding mechanics in fishes. One such area that has contributed substantially is hydrodynamics. Advances have emerged in both experimental and modeling techniques.

Visualization of Fluid Movement

The flow of water generated during suction feeding can be quantified using DPIV, which is a modern computational technique to visualize the flow of fluids. This technique has recently been utilized to study the flow generated by suction in centrarchid fishes and sharks. The speed of the water being sucked into the fish's mouth is highest at the mouth aperture and decays rapidly moving away from the fish's mouth. Thus suction feeding is most effective when the mouth of the predator is in very close proximity to the prey item.

Direct measurements of fluid flow permit the calculation of the ingested volume, which is strongly correlated with mouth size. Because the flow of water is unidirectional in fishes (flows in the mouth and out the opercular cavity), the amount of water ingested can exceed the size of buccal cavity. In addition to the size, it can also be calculated the shape of the IVW, which is strongly influenced by swimming speed.

Swimming during suction feeding can alter the hydrodynamics of suction feeding. Direct measurements from bluegill sunfish and largemouth bass indicate that, as swimming speed increases, the ingested volume becomes increasingly narrow and elongate

Ecomorphology of Feeding

The relationships between morphology and ecology, and ultimately performance, have been well studied in fishes. In particular, several studies have focused on the relationship between diet and structural aspects of the mouth. Recent work has provided a functional foundation for such studies, making the link between morphology and ecology more evident. For example, mouth size is strongly correlated with prey size such that species with larger mouths are able to ingest larger prey items.

Jaw lever mechanics are commonly studied in relation to ecology. One can measure the closing and opening in- and out-levers of the jaw, and then relate that to the diet or feeding style of the predator. In coral reef fishes of the Caribbean, species that use their oral jaws to grip, manipulate, bite, shred, or crush prey have very high jaw closing lever ratios. In contrast, those species that exhibit ram- or suction-based strike tactics exhibit much lower jaw closing ratios. Higher closing ratios indicate the ability to more efficiently transmit the force generated by the muscles to the tooth surface. Jaw opening ratios do not show an obvious pattern like the jaw closing ratios.

The ecomorphology of feeding in cottid fishes (sculpins) has also been examined in detail. The focus has primarily been on the species of the Northeastern Pacific. In these studies, a larger mouth was linked to feeding on more elusive prey (that is, fishes, shrimp, mysids, and octopods). In addition to the general ecomorphological approach, functional studies have also been used to test whether the morphology exhibited by some species enhances their performance in an ecological context. Larger-mouthed species of sculpins exhibit significantly higher capture success on evasive shrimp than did smaller-mouthed species. This functional link is critical for providing a mechanism underlying the correlation between morphology and ecology.

Jaw Protrusion

The flow of water generated by suction is limited to a narrow range outside of the fish's mouth. Thus, being able to get extremely close to a prey item is necessary for effective suction feeding. Jaw protrusion is one way in which the mouth can be extended closer to the prey without having to move the entire body. Taken another way, the predator could initiate a strike at a farther distance from the prey.

The ability of fishes to protrude their upper jaw has been cited as a key innovation that facilitated the trophic diversity among fishes. The premaxilla is rotated or pushed into an extended position, and multiple paths to jaw protrusion have evolved. The upper jaw movement can involve anterior, anterodorsal, or anteroventral aspects, and the lower jaw or mandible can pivot around a ball and socket articulation.

Recent work has revealed that jaw protrusion enhances forces exerted on a prey item. Although this had been suggested for quite some time, empirical data exploring the functional ramifications of jaw protrusion are limited. Utilizing a framework based on data from bluegill sunfish, a recent study utilized empirical data and simulations to calculate the contribution of jaw protrusion to the forces exerted on prey. The observed speed of jaw protrusion was positively correlated with the peak measured force on the prey. Ultimately, the independent source of acceleration provided by jaw protrusion can increase the total force on the prey by up to 35 per cent.

Locomotion and Feeding

For several reasons, locomotion is critical for most fishes to successfully capture prey. Locomotor performance (namely, maximum swimming speed and maximum acceleration) will be important for overtaking an evasive or elusive prey item. In addition, controlling the locomotor system (namely, stability and positioning) will play a significant role in the timing and accuracy of a strike, which will ultimately determine capture success.

Forward Movement of the Fish (Ram)

Several species of fishes capture prey without generating suction or manipulating the prey. Ram feeding occurs when a fish swims over a prey item with its mouth open. Examples of fishes that employ ram include tunas and whale sharks.

Speed of Forward Movement of the Fish

Although ram is often a component of a feeding event, the relative contribution can vary considerably. One ramification of increased ram speed is a decreased capture success. Among other things, swimming fast during prey capture decreases the time that the predator has to adjust its position relative to the prey. This can result in poor aiming and/or incorrect timing of mouth opening. One critical aspect of suction feeding that is altered with ram speed is the shape of the IVW. As a fish swims faster during feeding, the resulting IVW will be increasingly elongate and narrow. This is important with respect to positioning of the prey in front of the predator's mouth. If the prey item is located farther away from the predator, an elongated IVW will facilitate prey capture.

Manipulation

In addition to suction and ram, fishes can obtain prey by physically contacting it and bringing it into the mouth, that is, biting. Biting is relatively common among shark and teleost fishes. Every species of heterodontid sharks is ecologically and functionally specialized for durophagy, which involves the consumption of hard prey, such as, mollusks or crabs. In many cases, the oral jaws of fishes are used to bite a prey item that is located on the substratum. Furthermore, some fishes will bite other animals (namely, specialized for eating scales) and some will bite off small pieces of plants. Jawless fishes also use manipulation in order to obtain food.

Feeding in Jawless Fishes

Hagfishes and lampreys feed in a complex and very different way compared with other fishes. The feeding system includes posteriorly directed keratinous teeth that are attached to dental plates, a paired series of cartilages. The dental plates are moved in and out of the mouth during feeding via retractor and protractor muscles. The movement of the dental plate acts in a similar way to a fixed pulley. This system appears to have both advantages and disadvantages relative to fishes with jaws. The pully system allows maximal force transmission, but does not permit speed amplification. Speed amplification is likely a key aspect of jaws that has facilitated the success of jawed fishes.

Pectoral Fins

Strike accuracy is necessary for successful prey capture in fishes, especially those species that rely heavily on suction. In acanthomorph fishes, the pectoral fins are well suited for braking without inducing any unsteady maneuvers because of their location relative to the fish's center of mass. For these fishes, the reaction force generated by the pectoral fins during braking goes through the center of mass, thus maintaining the fish's position. Many fishes brake during prey capture, but the reason is still in question. One explanation is that braking reduces the approach speed, which will allow the predator to adjust its position relative to the prey. Thus braking may increase capture success. Another possible benefit to braking during prey capture is that it puts the predator in a position to quickly maneuver and follow the prey if it escapes. Finally, braking might prevent a collision with the substrate (if the prey is located close to structure) which will ultimately prevent injury.

Fish Feeding Behavior Utilized for Capture Fisheries

Some of the feeding behaviors of fish (namely, ram feeding and suction feeding) have been profitability utilized in capture fisheries.

There are several ways a fish may take a prey or bait, such as, an attack action, where the fish approaches its prey rapidly and bites into it, or sucking action, where it approaches slowly and cautiously and then suck it in. But in either case there is a common process to the act of taking in a prey or bait. The mouth is opened as the fish approaches the prey and then sucks it in along with the water around it with the gill covers open wide on both sides. Once inside the mouth the water is forced out through the gills and the prey or bait is taken into the stomach cavity.

Ram Feeding

The forward movement of fish with open mouth (in tunas and sharks) have been taken advantage of catching fishes with troll line. In trolling (the most active type of angling operation) the vital parts involved in the fishing process include, (i) the location and persuit of a school of fish, (ii) setting the proper trolling depth for fishing line and (iii) the performance of the fishing hooks and bait.

In surface trolling, the lines and hooks are towed horizontally, near the water surface. The fish caught by this method are pscivorous species with a large migrating range, such as middle to large sized tuna, marlin, skipjack, Spanish mackerel, yellowtail. Effective lures have been developed which attract the fish by the quick "swimming" motion they stimulate as they move through the water. The fishes with ram feeding behavior chase the lure with open mouth and accelerated speed and gulp the lure and being caught in the line.

Both live baits, such as, squid pieces and lures such as, feather fly are used in trolling. It is extremely important that the bait be one that appeals strongly to the fish and stimulates its feeding instincts by (a) moving through the water with a quick "swimming" motion similar to that of a real fish, (b) that it provides strong visual stimulus by means of bright colors and intermittent flashing of reflected light and (c) that it has a slight sound as it swims like that of a real group of migrating fish.

Sucking Action

The fishing hook is an implement that takes advantages of their sucking action to catch the fish. First the hook with bait attaches to the bend portion hangs in a balanced position in the water. Then when the fish sucks in the bait and hook, the gravity of the shank causes a shift in that balance that sends the point into a circular motion with the eye as the pivot point, causing it to stick.into the upper jaw of the fish.

The shape of the fishing hooks varies with the type of fish intended to catch, but the basic shapes are angular, round and elongated types. In order to prevent the captured fish from dropping out of the hook, most hooks have a barb. In some cases, when it is necessary to remove the fish from the hook quickly, as in case of pole-and-line skipjack fishing, hooks without barb is used. The hook with point bent inward, is the most suitable for the use with bottom long line as it does not enter the fish's body deeply and thus allow the fish to remain alive for sometime after swallowing the hook

Lure hooks are used for schools of fish with strong feeding behavior. The hook is hidden by a featherfly, is jigged back and forth along the water surface like a small fish jumping from the water.

Fish Behavior on Locomotion and Distribution

Oceanic tunas are mostly large in size and found in the oceanic habitats. In the tropical oceans, they consists of skipjack tuna (*Katsuwonus pelamis*), yellowfin tuna (*Thunnus albacares*), bigeye tuna (*Thunnus obesus*0, albacore (*Thunnus alalunga*) and southern bluefin tuna (*Thunnus maccoyii*).

Skipjack tuna (generally weighing less than 5 kg) are found throughout all intertropical waters. This species are found throughout the Indian Ocean from as far as 40-45 degree south in the west as well as off South Australia.

Yellowfin tuna is widely distributed throughout the Indian Ocean. However, it is not found south off Australia, where the influence of Antartic waters are strongly felt. It is less abundant in the north of the Arabian Sea (Gulf of Oman) and in the south of the Gulf of Aden.

Bigeye tuna, like the yellowfin, are present throughout the intertropical zone. They are also found in bordering areas where the yellowfin are scare, such as, the Arabian Sea with its low oxygen levels, and also in sub-tropical areas where temperature are lower. Bigeye is found in greater depths, its vertical distribution seems to be closely linked to the thermocline. Young bigeye schools are found frequently below drifting wreckage in 50-100 meter of water.

Young albacore are found in subtropical areas and in shallow waters of 15-20 degree Celsius; and the older ones towards the low latitudes in deep waters in temperatures of 14-25 degree Celsius. They may be found on the surface occasionally. Few albacores are found north of 10 degree north in the Indian Ocean and south of approximately 35-40 degree south in the Western Indian Ocean and 25-30 degree south in the Eastern Indian Ocean. Some albacore may be found south of Australia as far as 40 degree south.

Southern bluefin tuna is found only in the southern hemisphere. It is widely distributed throughout the three oceans from 30 to 50 degree south and in relatively cold waters (5-10 degree Celsius). However, this species can be found as far as 10 degree south in the Indian Ocean, south of Indonesia, where they spawn. Adult southern bluefin tuna are caught south of Tasmania from June to September and from May to August off South Africa.

Traits

The tunas have evolved a counter-current heat-exchange mechanism for conserving metabolic heat and raising body temperature. The body temperature of

Euthynnus alletteratus was found to be 11.3 degree Celsius higher than the surrounding water temperature (the body temperature was 31.2 degree Celsius, while the surface water temperature was 19.9 degree Celsius). The excess temperature (defined as the body temperature minus ambient water temperature) have been found to be reduced in captivity. To maintain the core body temperature warmer than ambient temperature of sea water, tuna uses centrally located vascular heat-exchange mechanism.

The warm central core of tuna is related to the distribution of red muscle. The warmest body temperature occurred along the vertebral column, between the first and second dorsal fin. The heat exchange mechanism is composed of dorsal aorta, the posterior cardinal vein and a large vertical rete. The dorsal aorta is embedded in the posterior cardinal vein, and is thus completely bathed in venous blood. The oxygen carrying capacity of haenoglobin is unaffected by changing blood temperature.

The principal advantage of a high and fairly constant body temperature is facilitation of continuous swimming by increasing the frequency of muscle contractions, thus increasing available swimming power. It was observed that warm blooded fishes achieve a marked independence from environmental temperature that permits them to make rapid horizontal and vertical migrations without the necessity of thermal acclimation.

Aspects of gas exchange (*E. affinis*), the perfusion rate of the gills ranged from 1.55 to 1.88 liter/minute per kg. The head loss in cm of water average 1.8. The calculated resistance after flow offered by the gills averaged 1.03 cm of water per minute per kg per liter and ranged from 0.90 to 1.20. Oxygen uptake ranged from 400 to 643 mg per hour per kg. The calculated utilization (fraction of oxygen removed from the water) ranged from 0.69 to 0.95.

Tuna possess a pineal apparatus, function as a light receptor to control phototactic movement of fish. Bluefin tuna's pineal organ has sensory cells possessing the structural characteristics of vertebrate retinal photoreceptor cells. The pineal organ in Bluefin tuna serves as photoreceptor, which receives external light stimulus.

The swimming speed of tuna depend on the adaptive radiation in the morphology, especially of the gas bladder and pectoral fins, which together with swimming speed contribute to the mechanism by which tunas maintain hydrostatic equilibrium and problems associated with maintaining hydrostatic equilibrium by large body size.

In attacking maneuver, the tuna swam downward to their prey with a great burst of speed and in the instant before reaching bottom, they turned sharply parallel to the bottom, stirring up clouds of sediment. Tuna took their prey at this point, or a fraction of a second later and the fish's jaws could be heard snapping audibly.

The gregarious nature of tuna causes them to swim in schools, which can be observed from the surface. This behavior is considered generally as being a protection measure and can take different forms according to the area or even according to the hour of the day.

Surface schools can be spotted by observations, swim freely or aggregate with flotsam. Free swimming schools, swim close to the surface or sub-surface and are accompanied by birds. To qualify a fish school according to the behavior of the individuals comprise of;

☆ Breezer – the presence of tuna indicated by a rippling on the surface, caused by fish swimming sub-surface in the same direction. This situation signals very often the presence of large schools.

☆ Finner – the fish do not surface entirely, only dorsal fins can be seen surfacing from time to time.

☆ Jumper – single fish jump out of the water and dive head first (fish behaving in this way are generally thought to have momentarily lost contact with the rest of the school).

☆ Smoker – the fish collectively jump out of the water, generating a choppy sea. This behavior is characteristic of mixed schools, made up of small tunas.

☆ Boiler or foamer – This term is used when the preceding situation becomes highly accentuated and is usually caused by large fish preying on anchovies or euphausiids. Even from a distance the fish can be seen to jump in a disorderly fashion causing the water to foam.

Schools are also aggregated around floating objects (flotsam, wreckage) or anchored objects (FADs) and mammals (whales, dolphins) or sharks. These schools may be or may not be accompanied by birds and may be composed of one or more species.

Deep sea schools may be free swimming or may aggregate with floating objects and can be detected acoustically.

1. Only detected below flotsam, compact in form, comprised of adult skipjack mixed with small yellowfin and bigeye.

2. Both free swimming and aggregated schools, comprise of two parts, one above the other. The upper one is compact and lower one is longer. Sometimes two sections are separated, made up of tuna mixed in size and species. The smaller individuals form the upper stratum in a compact formation and the larger ones, the second deeper ones.

3. Typical free swimming formation comprised of large fish may be occupying a thick layer of water and registering an oblong diamond shape on the echo-sounder.

Bigeye tuna is epipelagic and mesopelagic in ocean waters, occurring from the surface to about 250 m depth. Temperature and thermocline depth seem to be the main environmental factors governing the vertical and horizontal distribution of bigeye tuna. Water temperatures in which the species has been found range from 13 to 29 degree Celsius, but the optimum range lies between 17 and 22 degree Celsius. This coincides with the temperature range of the permanent thermocline. In fact, the tropical western and central Pacific, major concentrations of *T. obesus* are associated with the thermocline rather than with the surface phytoplankton maximum. For this reason, variation in occurrence of the species is closely related to seasonal and climatic changes in surface temperature and thermocline.

Juveniles and small adults of bigeye tuna school at the surface in mono-species groups or together with yellowfin tuna and/or skipjack schools may be associated with floating objects.

Schooling of yellowfin tuna occurs more commonly in near surface waters, Primarily by size, either in monospecific or multi-species groups. Larger fish greater than 85 cm fork length, frequently school with porpoises. Association with floating debris and other objects is also observed.

Skipjack tuna, an epipelagic, oceanic species with adults distributed within 15 degree Celsius isotherms, overall temperature range of occurrence is 14.7 to 30 degree Celsius, while larvae are mostly restricted in water with surface temperature at least 25 degree Celsius. Aggregation of this species tend to be associated with convergences, boundaries between cold and warm water masses, upwelling and other hydrological discontinuities. Depth distribution ranges from surface to about 260 m during the day, but limited to near surface water at night.

Floating Behavior of Fish

Many fish, such as the cod and the perch, bring themselves to neutral buoyancy by having gas space within themselves. This space amounts usually to about 5 per cent of the total volume of a marine fish, and gives a lift that just balances the weight in sea water of the other tissues. The swimbladder wall is not rigid, so that if a fish swims upwards or downwards in the sea and so changes the external pressure acting on it, the gas in the swimbladder either expands or is compressed and this changes the animal's buoyancy. In response to longer-lasting changes, fish can secrete more gas into the swimbladder or reabsorb gas from the swimbladder so as to restore neutral buoyancy. The swimbladder is used down to astonishing depths, to 2000 m and probably down to 4500 m, at this later depth, the pressure will be 450 atmospheres, and the density of the gas itself will demand that the volume be greatly increased if it is to provide neutral buoyancy.

It has been found that whereas the gas from the swimbladders of fish living near the surface of the sea often contain less oxygen than air, the proportion of oxygen increases with depth and gas taken from fish living at appreciable depths is mostly oxygen. Some swimbladders must be capable of secreting oxygen against very steep pressure-gradients. The gas is secreted into the swimbladder from a special tissue, known as gas gland. This secretion seems to be intimately linked with the structure called the rete mirabile in which arterioles going towards the gas gland break up into capillaries that come into intimate contact with corresponding venous capillaries arising from the gas gland. These capillaries are longest known in nature and deeper the fish lives, longer are the capillaries. The secretion of gas into the swimbladder is clearly the principal problem, for the mechanism of gas reabsorption are easy to understand. Some of the fish have a valved duct leading to the oesophagus through which they can allow gas to escape. Others have structure called the oval, in which some region of the swimbladder wall can be either exposed to or excluded from the gases in the swimbladder by the action of a ring muscle round its perimeter. This region is served by a blood supply without a rete, and gas can thereby be carried away from the swimbladder when the oval is open.

Hydrodynamic Behavior of Fishes

Fishes in still water, sometimes cease making swimming movement, glides forward by their momentum, as if it was rigid, but gradually moves more slowly. This retardation process can be determined by the rate of loss of momentum and hence the force with which the water resists the fish's motion.

This resistance of the water is much the same for a live fish as for a wooden object of the same shape and size. Adimittedly, for most fishes, this shape is streamlined and gives rather a low resistance. However, a fish can maintain a steady speed only if it can move its body and fins, so as to produce a net forward force or thrust exactly balancing the resistance of the water.

In the hundreds of millions of years during which fishes have evolved, aspects of swimming have in many environments had particular importance for survival. In the open ocean, the speed may be one such quality. A great many marine fishes live by capturing and eating other organisms, and probably most die either from starvation or by being themselves captured and eaten. Small improvements in speed can reduce the chance of premature death through either of these causes. In consequence, fishes have acquired some remarkable capabilities; for example, a tuna fish about the size of a man can swim ten times as fast as an Olympic champion.

Another quality that has survival value is swimming efficiency. Without this,the fish would be too rapidly use up the supply of energy derived from food while moving around to find its next meal. A swimminf fish, which produces a thrust balancing the resistance of the water, is doing work, "useful work" in the engineer's sense, at a rate equal to thrust multiplied by speed. At the same time, its body and fin movements may be wasting energy by churning up the water behind it into a turbulent wake. For efficiency it is important that the rate at which energy is wasted in making this eddying wake is as small a fraction as possible of the rate at which useful work is done.

Fishes representing an earlier line of vertebrate development, do not have a bony skeleton, but a cartilaginous one. Fishes with perfected bony skeleton, collectively called teleosts, possess in general a very effective hydrostatic organ, in the shape of a bladder full of gas, the so-called swimbladder and they can control the quantity of gas in the bladder in such a way that their weight is exactly balanced by their buoyancy. The teleosts, can, when they choose, cease swimming movements altogether without any resulting tendency to sink or rise.

Among the swimming methods, one appears to have been the swimming mechanism of the earliest fishes. It depends upon a transverse wave, or side to side undulation, passing down the body. Such an undulatory mode of swimming is found quite commonly among the invertebrates, with the wave passing back from head to tail and increasing somewhat in amplitude as it does so. It occurs in the more efficient form in the vertebrates, including both the more primitive jawless lampreys and the fishes proper. For vertebrates, this lateral undulation is much more effective because they have laterally compressed tails which greatly improve the efficiency of the swimming method.

Such lateral undulation is used by dogfishes and also by some sharks as well as by sturgeons and by lung-fishes. It is also used by certain group of teleosts, notably the eels with their highly extended shape. Fishes of generally similar shape like the ribbon-fishes and unicorn-fishes, as well as by various other fishes including cod.

With teleosts, in which weight is exactly balanced by buoyancy, no special movements are required for maintaining their vertical position, and therefore the undulatory mode is particularly easy to follow. The common eel, *Anguilla vulgaris* having laterally compressed tail, reduced body section, long continuous fins present dorsally and ventrally helps in maintaining the total body depth, although the lateral thickness is enormously reduced. At the posterior end, the tail practically becomes a vertical edge, which an aerodynamicist would call a "trailing edge".

In an eel swimming in a tank, the wave can be seen by progress backwards towards the tail. Positions of greatest curvature appear gradually further back. The speed of the wave down the body is always found to be greater than the resulting fish-speed through the water, so that even relative to the water the wave passes backwards. The amplitude of lateral motion increases as the wave passes from head to tail. This is the typical anguilliform mode of swimming.

Though fish like the cod does not have continuous dorsal and ventral fins, but has three of each, with only short gaps between them. These short gaps become filled up with vortex sheets and behave mechanically almost like a solid fin, while swimming, according to the theory of flow around slender bodies, so that the anguilliform motion of the cod becomes effectively very similar in its mechanics to that of the eel.

In carangiform swimming, the front part of the fish has lost its flexibility and the undulation is confined almost entirely to the rear half, or even third of the body length. A wave that passes backwards can still be discerned, but it is a wave whose amplitude increases rather fast from almost zero at the mid-point of the fish's length to a large value at the tail. As before, the trailing edge's motion lags behind that of front section although now there is almost no motion further forward still. This modified undulatory motion confined to the neighbourhood of the tail is known as carangiform motion.

It is known from engineering experience that this rapid acceleration produces a better result because it keeps the kinetic energy of the water's motion down to the half *mw* square value, whereas slower acceleration gives time for water motions with extra kinetic energy to appear, due to eddies shed from the tops and bottoms of fine cross-sections.

Carangiform motion is well developed in the salmon family, *Salmo salar*. The front half of the body is not at all flexible, but over the rear half there is a rapid increase in wave amplitude to the large value it reaches at the caudal fin. The fishes, using carangiform method swimming include the horse-mackerels and jacks of the family Carangidae, as well as perches, red mullets, barracuda and red cardinal fishes.

The fins in the fish's plane of symmetry, the dorsal, ventral and above all, caudal fins are important for propulsion and for stability against side to side movements.

However, paired fins off the plane of symmetry also exist. These are the pelvic fins and more important, the pectoral fins (just behind the head). The paired fins give stability against heaving and pitching motions. Fins in general are important, too, in enabling fishes to make fine adjustments of their position in the water, and in stopping and starting. Fishes are helped to make a rapid start from rest because muscle can exert about four times as much power for a brief period as it can continuously. Fishes therefore can start quickly by making normal swimming motions with extra force. On the other hand, stopping is not so easy because the streamlined shape of the fish enables it to glide forward a considerable distance under its own momentum. For a sudden stop, however, fishes have learnt to use their pectoral fins so as greatly to increase water resistance, much as an aircraft pilot uses his airbrakes.

Fishes which have succeeded best in the struggle for existence in the surface waters of the deep ocean, and in doing so have become the fastest fishes of all. They have achieved this by a major change in the shape of the caudal fin, what an aerodynamicist would call an increase in its aspect ratio, defined as depth square/ surface area. There is already an aeronautical appearance about the herring fin, which closely resembles a pair of highly sweptback wings; increasing the aspect ratio can be regarded as making them not too sweptback.

Thrust can be increased if there is an increase in the virtual mass at the trailing edge, and this mass in turn is proportional to the square of the depth of the caudal fin. However, if the caudal fin were made too big, the resistance as its large area was dragged through the water would become excessive, so that speed might not much improve. This indicates the importance of the aspect ratio, which can be increased by reducing the sweepback well below. As a result, thrust is raised without much increase in resistance, and the fish can go faster.

The tunas, striped marlin, the sailfish and swordfish, all are with different ends, but with broadly similar lunate tails. These are all outstandingly fast, active fishes, constantly on move, to such an extent that they have gradually lost all pumping apparatus for bringing water in contact with their gills. It is no longer necessary because their unceasing motion forces water through their mouths and out of their gill slits to a quite sufficient extent.

There is something about the lunate tail that especially fits it for high speed marine proultion. The method is strictly carangiform, all the propulsive mechanism being in the tail, and the reduction in cross section depth just before the caudal fin is even more extreme than in the salmon. All thse fishes possess exceptionally powerful musculature for moving the tail at very high frequency (as much as 10 Hz), and the body temperature is found to be unusually high (almost 30 degree Celsius).

All the lunate tails show a strong similarity to certain configurations of aeroplane wings; they have good "aerofoil sections", with a nice blunt leading edge. It is in fact possible to analyse the thrust developed by a fish's lunate tail due to its carangiform motion by just the methods used to find the forces sustained by aircraft wings when pitching and heaving.

The lunate of tuna, as the fish moves to the left, shows the eddy or vortex cast off at each extreme of the tail's movement and depicts the backward stream of water that these vortices generate between them, indicating that this way of obtaining thrust has something in common with jet propulsion. The character of the vortices at roughly the level of the fish's nose. In three dimensions, taking into account the curved shape of the lunate tail in the vertical, it can be regarded as a horizontal section of a sequence of vertical vortex rings pushed backward diagonally, alternately to the right and to the left by the fish. It is well known that smoke rings (which are vortex rings in air) have enormous momentum, and the lunate tail may be so effective propulsively because it can especially well give thrust by the reaction of these backward moving vortex rings, that have so much momentum in relation to their energy.

Fishes heavier than water are not so outstandingly fast or efficient in their swimming, because their weight is not exactly balanced by the buoyancy of a controlled volume of swimbladder gas.These include most of the sharks. The slower sharks and most of the dogfishes have a specially shaped asymmetric (heterocercal) tail, which support their excess weight while swimming. Essencially, the heterocercal tail consists of a single large sweptback wing above and a much smaller one below. As the large wing moves from side to side, it produces thrust at right angles to itself, which therefore, contains a certain vertically upward component, or lift. At the same time the pectoral fin is well developed, and set an angle of incidence to carry the main part of the load, just as does an aeroplane wing. The weight of an aeroplane can be balanced in a stable fashion only if wing lift ahead of the center of gravity takes part of the load while tail lift behind takes the rest. With the shark the arrangement is the same, the pectoral fin is like the wing, and the anguilliform motion of the body and the heterocercal tail gives not only the thrust, but also the necessary tail lift.

The same arrangement is found in the sturgeon, which has the same problem of weight support. All these animals, as a result, have just the dynamical properties of an aeroplane and they can, for example, gracefully "loop the loop".

In skates, rays and allied fishes, there is still more pronounced development of the pectoral fin, initially for weight support, but increasingly contributing to propulsion. Some like the so-called guitar-fishes, still have effective caudal fins.

The bottom living skates and rays, have in fact become almost all pectoral fin, the two fins together approximate to a square, with an extremely thin body and tail down the diagonal. They swim very beautifully by passing backwards over the pectoral fins an undulation very like the basic undulatory mode with which it was started, but now with up-and-down motions instead of side to side. These fish live near the bottom and when not actively swimming, their wing-like shape, with aspect ratio of about 2, enables them to achieve a conveniently low angle of glide down to the bottom.

On the other hand, eagle rays have developed a mode of swimming even closer to the flight of a bird or devil rays swim in a rather similar manner. Their pectoral fins have developed into wings having an aspect ratio of around 4 and much less than a whole wave length of undulation is visible at any one time. They use in fact a strong downstroke and a rather more feathered upstroke, like many birds.

The bottom living teleosts, like plaice have lost their swimbladder, so that when they are inactive they glide down on to the bottom, where their excellent protective coloring gives them some security. However, when very young, they are fishes of normal shape, swimming in anguilliform motion, and it is only at a certain age that they turn over on their sides and the lower eye comes round to the top. Then their lateral anguilliform motion, turned through 90 degree, becomes an up-and-down one, and their final shape and mode of swimming are not unlike those of a skate, although the skate reached this condition quite differently, by an enormous development of the pectoral fin.

References

Biswas, K.P. *Marine Biology*, Daya Publishing House, A Division of Astral Internanational Pvt. Ltd., New Delhi.

Biswas, K. P. Oceanic Tuna, a fish of unique trait and international commodity; *Soc. for Indian Ocean Studies*, Vol. 21, No. 1, 2013

Biswas, K. P. On the selection of effective impulse frequencies for specific movements of fish in an electric field; *Fishery Technology*, Vol. VIII, No.2, 1971.

Biswas, K. P. Determination of threshold current densities for different reactions of fish using a Pantostat; *Fishery Technology*, Vol.VIII, No.1, 1971.

Biswas, K. P. Studies on threshold current densities for convulsions of fish using a square wave stimulator, *Fishery Technology*, Vol.VIII, No.2, 1971.

Biswas, K. P. On the behavior of marine crustaceans in an electric field of alternating current, *Fishery Technology*, Vol.VIII, No.1, 1971.

Biswas, K.P. and S.P.Karmarkar. Reaction of fishes to the underwater AC field, *Fishery Technology*, Vol. XIII, No.1, 1976.

Biswas, K. P. On the reaction of marine fishes in interrupted AC of 50 Hz, *Fishery Technology*, Vol. VIII, No.2, 1971.

Biswas, K. P. and S.P. Karmarkar. Effect of electric stimulation on heart beat and body muscle in fish *Fishery Technology*, Vol. 16, No.2, 1979.

Burnet, A.M.R Studies on the ecology of the New Zealand freshwater eels, Fisheries Research Laboratory, Marine Department, Wellington, NZ, 1952.

Biswas, K.P. Preliminary observations on the effect of electric seine on fish catch, *Fishery Technology*, Vol. VII, No.2, 1970.

Burrows, R.E. Diversion of adult salmon by an electric field, Sp. Sci. Rep. No.246, Washington DC, 1957.

Biswas, K.P. Studies on the effect of electrical energy on certain aquatic organisms, A thesis submitted to the University of Bombay for the Degree of Master of Science in Zoology, 1974.

Biswas, K.P. Further studies on the effect of electrical energy on certain aquatic organisms, A thesis submitted to the University of Bombay for the Degree of Doctor of Philosophy in Zoology, 1977.

Cowx, I.G. and P. Lamarque. *Fishing with electricity, Application in Freshwater Fisheries Management*, Fishing News Books, 1990.

Chmielewski, A. Study of reactions and behavior of fish in Heterogeneous Field with Single and Multiphase current, *Fish Screen and Guides*, 1964.

Denzer, H.W. Die Klektrofischerei, *Stuttgart*, 1956.

Evans, David H, *The Physiology of Fishes*, CRC Press, Boca Raton, New York, 1997.

Edwards, J.L. and J.D.Higgins. The effects of electric currents on fish, Game and Fish Division, Deptt. Of Natural Resources, Georgia and Sport Fishing Institute, Washington DC, 1973.

Funk, J.L. Wider application of the electrical method of collecting fish, *Trans. American Fisheries Society*.

Godfrey, H. Mortalities among developing trout and salmon ova following shock by Direct Current electrical fishing gear, *J.Fish.Res.Bd.Canada*, 14(2), 1952.

Godfrey, H. Catches of fish in New Brunswick streams by Direct Current electro-fishing, *The Canadian Fish Culturist*, No. 19, 1956.

Higham, T.E. Feeding mechanics, *Encyclopedia of Fish Physiology*, Academic Press, San Diego, 2011.

Halsband, E. Untersuchungen uber den Einfluss Verschiedener Stromarten auf den Stoffwechsel der Fische, 1954.

Halsband, E. and I. Veranderungen des blutbildes, von Fischen infolge toxischer schaden 1954.

Halsband, E. Veranderung der Erythrozyten bei Regenbogenforelle (*Salmo gairdnerii*) und Karpfen (*Cyprinus carpio*) durch einwirkung von Gleichstrom, *Arch. Fischerreiwiss, Braunschweig*, Vol. X, No.3, 1960.

Halsband, E Die Veranderung der Blutzellen von fischen nach Einwirkung von Elektrischem Strom und Rontgenstrahlen, *Elektro Medizin, B und 7*, Nr 3, 1962.

Halsband, E. Die Beziehung zwischen Intensitat und Zietdaur des Reizes bei der Elektrischen Durchstromung von Fischen, *Archiv.fur Fischereiwissenschaft*, 7(1), 1956.

Hosl, A. Dangers and precautions in the electrical fishery, Electro-Beratung, Bayern, Munich, Germany.

Lissmann, H.W. Electric location by fishes, *J. Exp. Biol.* 35, 156, (1958).

Meyer-Waarden, P.F. *Electrical Fishing*, Food and Agriculture Organization of the United Nations, Rome, 1957.

Mc Lain, A.L. The control of upstream movement of fish with pulsed direct current, U.S. Deptt. of Interior, Fish and Wildlife Service, 1956.

Meyer-Waarden, P.F. Einfuhrung in die Elektrofischerei, Westliche Berliner Inge and Egon halsband Verlagsgesellschaft Heenemann KG, Berlin, 1965.

Newman, H.W. Effect of field polarity in guiding salmon fingerlings by electricity, Special Scientific Report No. 319, US Fish and Wildlife Service, Washington DC, 1959.

Patten, B.G. and The Bureau of Commercial Fisheries Type IV Electrofishing Shocker C.C. Gillaspie its characteristics and operation, Special Scientific Report, US Fish and Wildlife Service, Washington DC, 1966.

Pratt, V.S. A measure of the efficiency of alternating current and direct current Fish Shockers, Idaho Cooperative Wildlife Research Unit, University of Idaho, Mosco, *Trans. Am. Fish Soc.*Vol.81, 1951.

Pratt, V.S. Fish mortality caused by electrical shockers, *Trans.Am.Fish Soc.*, 1957.

Rommel, S.A.Jr. Oceanic electric field perception by American eels, *Science*, Vol. 176 J.D. Mc Cleave 1972.

Shetter, D.S. The electric shocker and its use in Michigan Streams, *Progressive Canadian Fish Culturist*, Vol. 16, No. 9, 1947.

Sloman, K.A, *Behavior and Physiology of Fish*, Academic Press, San Diego, 2006.

Smith, G.M.F. and Direct current electrical fishing apparatus, Report presented to P.F. Elson Fisheries Research Board of Canada, Fish Cult., No.9, 1950.

Wilson, R.W. and Balshine, S Sport Fisheries Care of Tropical Aquarium Fishes, Fishery Leaflet 411, Wildlife Div.of Fish Hatcheries, 1965.

Index